Marx Joyce
Hardy Austen
Abbott Montaigne Chesterton Emerson
Defoe Melville Cooper Hugo
Machiavelli Eliot
Haggard Grimm
Stoker Christie Molière
Carroll Montaigne
Wilde Maupassant Byron
Garnett Fitzgerald Engels Schiller
Goethe Einstein Hawthorne Kafka
Cotton Dostoyevsky Smith
Baum Kipling Doyle Hall
Leslie Dumas Henry Nietzsche Willis
Flaubert Turgenev Balzac
Stockton Vatsyayana Crane
Burroughs Verne
Curtis Tocqueville Gogol Vinci
Homer Widger Tolstoy Whitman Busch
Darwin Thoreau
Potter Freud Zola Twain Scott
Kant Jowett Lawrence Plato Harte
Stevenson Dickens Burton
Andersen Hesse
London Descartes Cervantes
Poe Aristotle Voltaire
Hale James Hastings Cooke
Bunner Shakespeare Irving
Richter Chambers
Doré da Benedict Alcott
Dante Shaw Pushkin
Swift Chekhov
Newton
Wodehouse

tredition

tredition was established in 2006 by Sandra Latusseck and Soenke Schulz. Based in Hamburg, Germany, tredition offers publishing solutions to authors and publishing houses, combined with worldwide distribution of printed and digital book content. tredition is uniquely positioned to enable authors and publishing houses to create books on their own terms and without conventional manufacturing risks.

For more information please visit: www.tredition.com

TREDITION CLASSICS

This book is part of the TREDITION CLASSICS series. The creators of this series are united by passion for literature and driven by the intention of making all public domain books available in printed format again - worldwide. Most TREDITION CLASSICS titles have been out of print and off the bookstore shelves for decades. At tredition we believe that a great book never goes out of style and that its value is eternal. Several mostly non-profit literature projects provide content to tredition. To support their good work, tredition donates a portion of the proceeds from each sold copy. As a reader of a TREDITION CLASSICS book, you support our mission to save many of the amazing works of world literature from oblivion. See all available books at www.tredition.com.

Project Gutenberg

The content for this book has been graciously provided by Project Gutenberg. Project Gutenberg is a non-profit organization founded by Michael Hart in 1971 at the University of Illinois. The mission of Project Gutenberg is simple: To encourage the creation and distribution of eBooks. Project Gutenberg is the first and largest collection of public domain eBooks.

Delineations of the Ox Tribe The Natural History of Bulls, Bisons, and Buffaloes. Exhibiting all the Known Species and the More Remarkable Varieties of the Genus Bos.

George Vasey

Imprint

This book is part of TREDITION CLASSICS

Author: George Vasey
Cover design: Buchgut, Berlin – Germany

Publisher: tredition GmbH, Hamburg - Germany
ISBN: 978-3-8472-1916-3

www.tredition.com
www.tredition.de

Copyright:
The content of this book is sourced from the public domain.

The intention of the TREDITION CLASSICS series is to make world literature in the public domain available in printed format. Literary enthusiasts and organizations, such as Project Gutenberg, worldwide have scanned and digitally edited the original texts. tredition has subsequently formatted and redesigned the content into a modern reading layout. Therefore, we cannot guarantee the exact reproduction of the original format of a particular historic edition. Please also note that no modifications have been made to the spelling, therefore it may differ from the orthography used today.

DELINEATIONS

OF

THE OX TRIBE.

THE

SANGA OR GALLA OX OF ABYSSINIA, *v.* p. 120.

DELINEATIONS

OF

THE OX TRIBE;

OR,

THE NATURAL HISTORY OF
BULLS, BISONS, AND BUFFALOES.

EXHIBITING

ALL THE KNOWN SPECIES
AND THE MORE REMARKABLE VARIETIES

OF

THE GENUS BOS.

BY GEORGE VASEY.

ILLUSTRATED BY 72 ENGRAVINGS ON WOOD, BY THE AUTHOR.

LONDON:
PUBLISHED BY G. BIGGS, 421, STRAND.
1851.

C. AND J. ADLARD, PRINTERS, BARTHOLOMEW CLOSE.

TO

WILLIAM YARRELL, Esq., F.L.S., F.Z.S.,

WHOSE SCIENTIFIC WORKS ON ZOOLOGY

PLACE HIM IN THE FIRST RANK OF NATURALISTS;

AND, MOREOVER,

WHOSE UNOSTENTATIOUS KINDNESS IN CONSULTING THE FEELINGS

AND ADVANCING THE INTERESTS OF OTHERS

IS RARELY EQUALLED,

This Volume is inscribed,

BY HIS SINCERE FRIEND AND ADMIRER,

THE AUTHOR.

[Pg ix]

PREFACE.

The primary object of the present work, is to give as correct and comprehensive a view of the animals composing the Ox Tribe, as the present state of our knowledge will admit, accompanied by authentic figures of all the known species and the more remarkable varieties.

Although this genus (comprising all those Ruminants called Buffaloes, Bisons, and Oxen generally,) is as distinct and well characterised as any other genus in the animal kingdom, yet the facts which are at present known respecting the various species which compose it, are not sufficiently numerous to enable the naturalist to divide them into sub-genera. This is abundantly proved by the unsuccessful result of those attempts which have already been made to arrange them into minor groups. Nor can we wonder at this want of success, when we consider that even many of the species usually regarded as distinct are by no means clearly defined.

The second object, therefore, of this treatise, is (by bringing into juxta-position all the most important facts concerning the various individual specimens which have been described, and by adding several other facts of importance [Pg x] which have not hitherto been noticed,) to enable the naturalist to define, more correctly than has yet been done, the peculiarities of each species.

A third object is to direct the attention of travellers more particularly to this subject; in order that, by their exertions, our information upon this class of animals may be rendered more complete.

A new and important feature in the present Monograph, is the introduction of a Table of the Number of Vertebræ, carefully constructed from an examination of the actual skeletons, by which will be seen at a glance the principal osteological differences of species which have hitherto been confounded with each other. A Table of the Periods of Gestation is likewise added, which presents some equally interesting results.

Several of the descriptions have been verified by a reference to the living animals, seven specimens of which are at present (1847) in the Gardens of the Zoological Society, Regent's Park. The several

Museums in the Metropolis have likewise been consulted with advantage.

I am indebted to Judge Furnam, of the United States, for some original information respecting the American Bison; and also to the late Mr. Cole, who was forty years park-keeper at Chillingham, for answers to several questions which I proposed to him on the subject of the Chillingham Cattle.

I beg to acknowledge my obligation to Mr. Catlin for kindly allowing me, not only to make extracts, but also [Pg xi] to copy some of the outlines from his 'Letters and Notes on the North American Indians,' a work which I do not hesitate to pronounce one of the most curious and interesting which the present century has produced,—whether we regard the graphic merits of its literary or pictorial department.

To Professor Owen and the Officers of the Royal College of Surgeons, to the Officers of the Zoological Society, and to the Officers of the Zoological Department of the British Museum, my sincere thanks are due for the kindness and promptness with which every information has been given, and every facility afforded to my inquiries and investigations.

With respect to the engraved figures, I have striven to produce correct delineations of form and texture, rather than to make pretty pictures by sacrificing truth and nature for the sake of ideal beauty and artistic effect.

I cannot conclude this Preface without expressing my thanks to Messrs. Adlard for the first-rate style in which this volume has been printed; particularly for the successful manner in which the impressions of the engravings have been produced, superior, in general, to India-proof impressions.

King Street, Camden Town;
May, 1851.

[Pg xii]

ADDENDUM.

PENNANT – BUFFON – GOLDSMITH – BEWICK – BINGLEY.

In addition to the critical remarks on the writings of others, on this subject, which the reader will find in the following pages, I have further to observe that, although Pennant and Buffon have held a very high character, for many years, as scientific naturalists, the portion of their works which treats of the *Genus Bos*, appears to have been the result of the most careless and superficial observation. With the exception of the facts and observations furnished by such men as Daubenton and Pallas, Buffon's works are little more than flimsy speculations. As to Pennant's history of the Ox Tribe, it is calculated rather to bewilder than to inform; it is, in fact, an incoherent mass of dubious statements, huddled together in a most inextricable confusion: as a piece of Natural History it is absolutely worse than nothing.

Goldsmith, Bewick, and Bingley, three of our most popular writers on Natural History, appear to have done little more than compile from Pennant and Buffon, and consequently are but little deserving of credit. These strictures apply exclusively to such portions of their works as relate to the Ox Tribe.

[Pg xiii]

INTRODUCTION.

Ruminantia is the term used by naturalists to designate those mammiferous quadrupeds which chew the cud; or, in other words, which swallow their food, in the first instance, with a very slight mastication, and afterwards regurgitate it, in order that it may undergo a second and more complete mastication: this second operation is called ruminating, or chewing the cud. The order of animals which possess this peculiarity, is divided into nine groups or genera, namely:—

>Camels.
>Llamas.
>Musks.
>Deer.
>Giraffes.
>Antelopes.
>Goats.
>Sheep.
>Oxen.

The last named forms the subject of the following pages, and is called, in zoological language, the *Genus Bos*, in popular language, the Ox Tribe.

One of the most interesting occupations which the wide field of Zoology offers to the naturalist, is the investigation of those remarkable adaptations of organs to functions, and of these again to the necessities and well-being of the entire animal. Nor does it in the least diminish our interest in the investigation of individual adaptations, or our admiration on becoming acquainted with them, that we know, *à priori*, this universal truth, that all the constituents of every organised body, be that organisation what it may, are invariably adapted, in the most perfect manner, to each other, and to the whole.

It is by a knowledge of this exact harmony in the animal economy, that the comparative anatomist can determine, with almost unerring precision, the genus, or even species of an animal, by an examination of any important part of its organisation, as the teeth, stomach, bones, or extremities. In some cases, a single bone, or even the fragment of a bone, is sufficient to convey an idea of the entire animal to which it belonged.

In illustration of this: — if the viscera of an animal are so organised as only to be fitted for the digestion of recent flesh, we find that the jaws are so contracted as to fit them for devouring prey; the claws for seizing and tearing it to pieces; the teeth for cutting and dividing its flesh; the entire system of the limbs, or organs of motion, for pursuing and overtaking it; and the organs of sense for discovering it at a distance. Moreover, the brain of the animal is also endowed with instincts sufficient for concealing itself, and for laying plans to catch its necessary prey.

Again, we are well aware that all *hoofed* animals must necessarily be herbivorous, or vegetable feeders, because they are possessed of no means of seizing prey. It is also evident, having no other use for their fore-legs than to support their bodies, that they have no [Pg 3] occasion for a shoulder so vigorously organised as that of carnivorous animals; owing to which they have no clavicles, and their shoulder-blades are proportionally narrow. Having also no occasion to turn their forearms, their radius is joined by ossification to the ulna, or is at least articulated by gynglymus with the humerus. Their food being entirely herbaceous, requires teeth with flat surfaces, on purpose to bruise the seeds and plants on which they feed. For this purpose, also, these surfaces require to be unequal, and are, consequently, composed of alternate perpendicular layers of enamel and softer bone. Teeth of this structure necessarily require horizontal motions to enable them to triturate, or grind down the herbaceous food; and accordingly the condyles of the jaw could not be formed into such confined joints as in the carnivorous animals, but must have a flattened form, correspondent to sockets in the temporal bones. The depressions, also, of the temporal bones, having smaller muscles to contain, are narrower and not so deep; and so on, throughout the whole organisation.

The digestive system of the ruminantia is more complicated in structure than that of any other class of animals; and, owing to this complexity, and the consequent difficulty of investigating it, its nature and functions have been less perfectly understood.

The stomach of the Manilla Buffalo, which will serve as an example of all the other species, is divided into four cavities or ventricles, which are usually (but improperly) considered as four distinct stomachs. [Pg 4]

The following figure represents the form, relative size, and position of these four cavities when detached from the animal, and fully inflated.

a. First cavity, called the paunch.
b. Second ditto, the honeycomb bag.
c. Third ditto, the many-plies.
d. Fourth ditto, the reed, or rennet.
e. A portion of the œsophagus, showing its connection with the stomach.
f. The pylorus, or opening into the intestines.

The interior of those cavities present some remarkable differences in point of structure, which, in the present work, can only be alluded to in a very general manner. For a particular account of the internal anatomy of these complicated organs, the reader is referred to the interesting work on 'Cattle,' by W. Youatt. [Pg 5]

The paunch is lined with a thick membrane, presenting numerous prominent and hard papillæ. The inner surface of the second cavity is very artificially divided into angular cells, giving it somewhat the appearance of honeycomb, whence its name "honeycomb-bag." The lining membrane of the third cavity forms numerous deep folds, lying upon each other like the leaves of a book, and beset with small hard tubercles. These folds vary in breadth in a regular alternate order, a narrow fold being placed between each of the broader ones. The fourth cavity is lined with a velvety mucous membrane disposed in longitudinal folds. It is this part of the stomach that furnishes the gastric juice, and, consequently, it is in this cavity that the proper digestion of the food takes place; it is here, also, that the milk taken by the calf is coagulated. The reed or fourth cavity of the calf's stomach retains its power of coagulating milk even after it has been taken from the animal. We have a familiar instance of its operation in the formation of curds and whey.

The first and second cavities (a and b) are placed parallel (or on a level) with each other; and the œsophagus (e) opens, almost equally, into them both. On each side of the termination of the œsophagus there is a muscular ridge projecting, so that the two together form a sort of groove or channel, which opens almost equally into the second and third cavities (b and c).

[As there has not been, as far as I am aware, any appropriate name given to this very remarkable part of the stomach of ruminants, I here take the liberty of [Pg 6] suggesting the term *Gastroduct*, by which epithet this muscular channel will be designated in the following pages.]

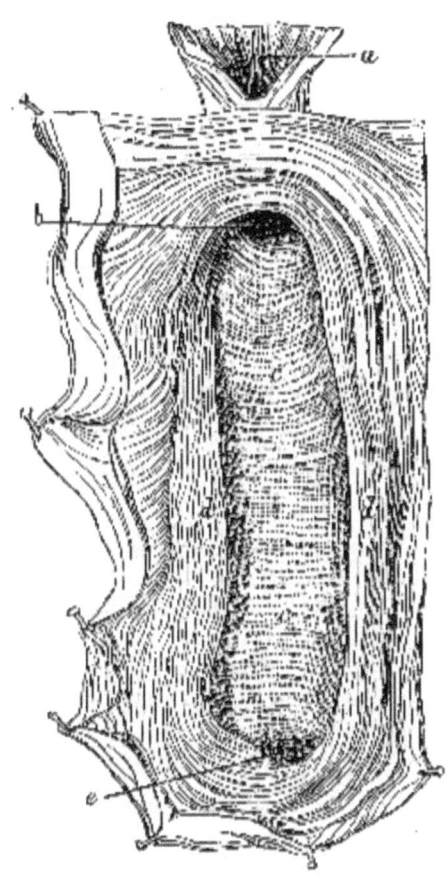

View of Gastro-duct, after Flourens.

a. A portion of the œsophagus cut open, showing the internal folds of the mucous membrane.
b. The opening of the œsophagus into the paunch.
c, c. The gastro-duct.
d, d. Muscular fibres passing completely round the edge of the gastro-duct, and forming a sort of sphincter.
e. The opening from the gastro-duct into the third cavity.

[Pg 7]

All these parts, namely, the œsophagus, the gastro-duct, and the first three cavities, not only communicate with each other, but they communicate by one common point, and that point is the gastro-duct. At the extremity of the third cavity, opposite to that at which the gastro-duct enters it, is an aperture which communicates immediately with the fourth cavity (*d*).

Such is a very brief description of the complicated stomach of the Ox Tribe. In what manner the food passes through this curious arrangement of cavities is a problem which has engaged the attention of naturalists from a very early period. A host of great men might be cited who have failed to solve it. The French physiologist, M. Flourens, by his recent experiments, has done more than any or all of his predecessors to give clearness and precision to this intricate subject.

The following is an abstract of the most important of his experiments:—

A sheep having been fed on fresh trefoil, was killed and opened immediately,—that is, before the process of rumination had commenced. He (M. Flourens) found the greatest part of this herb (easily recognised by its leaves, which were still almost entire,) in the paunch; but he also found a certain portion (*une partie notable*) of those leaves (in the same unmasticated state) in the honeycomb. In the other two cavities, (the many-plies and the reed,) there was absolutely none. [Pg 8]

M. Flourens repeated this experiment a great many times, with herbs of various kinds, and the result was constantly the same: from which it appears, that herbaceous food, on its first deglutition, enters into the honeycomb, as well as into the paunch; the proportion, however, being considerably greater into the paunch than into the honeycomb. It appears equally certain that, in the first swallowing, this kind of food *only* enters into the first two cavities, and never passes into the many-plies or the reed.

Having ascertained this fact with respect to *herbs*, he instituted a similar series of experiments, in which the animals were fed upon various kinds of *grain*,—rye, barley, wheat, oats, &c. The animals were killed and examined, as in the former experiments, immediately after being fed. He found the greater part of the grain unmasticated (*tout entier*) in the paunch; but, as in the case of the herbs, he also found a certain portion, in the same unmasticated state, in the honeycomb. Neither the many-plies nor the reed contained a single grain. He repeated these experiments many times, and always with the same result.

He then tried the effect of carrots cut into pieces, from half an inch to an inch in length; and in order that the animals might not chew them, he passed them into the pharynx by means of a tube. In one of these sheep he found all the morsels in the paunch; but, in the other two, some of the morsels were in the honeycomb, and some in the paunch. In all the three cases, [Pg 9] there was none either in the many-plies or in the reed.

He then proceeded to ascertain the effect of substances previously comminuted. He caused a certain quantity of carrots to be reduced to a kind of mash, with which he fed two sheep, and opened them immediately afterwards. He found the greatest part of this mash in the paunch and in the honeycomb; but he likewise found a certain portion in the many-plies and in the reed.

His next experiments were made upon plain fluids. It is the opinion of the generality of authors on this subject that fluids pass immediately and *entirely*, along the gastro-duct, into the third and fourth cavities. But, according to the experiments of M. Flourens, this is not the case. He found, by making artificial openings (*anus artificiel*) in the stomachs of various sheep, that, as the animals drank, the fluid came directly out at the opening, in whatever cavity it might have been made.

It is clear, then, that fluids pass, in part, into the first and second cavities, and, in part, into the third and fourth; and they pass as directly into the former as into the latter.

The following is the result of some experiments which M. Flourens made respecting the formation of the pellets.

In the first place, after the animal has swallowed a certain quantity of food the first time, successive pellets are formed of this food, which remount singly to the [Pg 10] mouth; secondly, there is a particular apparatus, which forms these pellets; and, thirdly, this apparatus consists of the two closed apertures (*ouvertures fermées*) of the many-plies, and of the œsophagus. Thus, the first two cavities, in contracting, push the aliments which they contain between the edges of the gastro-duct; and the gastro-duct, contracting in its turn, draws together the two openings of the many-plies and œsophagus; and these two openings, *closed* at this moment of their action, seize a portion of the food, detach it, and form it into a pellet.

The chief utility of rumination, as applicable to all the animals in which it takes place, and the final purpose of this wonderfully-complicated function in the animal economy, are still imperfectly known; what has been already suggested on these points is quite unsatisfactory. Perrault and others supposed that it contributed to the security of those animals, which are at once voracious and timid, by showing the necessity of their remaining long employed in chewing in an open pasture; but the Indian buffalo ruminates, although it does not fly even from the lion; and the wild goat dwells in Alpine countries, which are inaccessible to beasts of prey.

Whatever may be our ignorance of the cause or the object of rumination, it is certain that the nature of the food has a considerable influence in increasing or diminishing the necessity for the performance of that function. Thus, dry food requires to be entirely subjected to a second mastication, before it can pass into [Pg 11] the many-plies and reed; whilst a great portion of that which is moist and succulent passes readily into those cavities, on its first descent into the stomach.

It has already been shown by the illustration, (p. 4,) that the paunch is the largest of the four cavities; but this is not the case with the stomach of the young calf, which, while it continues to suck, does not ruminate; in this case the *reed*, which is the true digestive cavity, is actually larger than the other three taken together.

When the calf begins to feed upon solid food, then it begins to ruminate; and as the quantity of solid food is increased, so does the size of the paunch increase, until it attains its full dimensions. In this latter case, the *paunch* has become considerably larger than the other three cavities taken together.

A curious modification of an organ to adjust itself to the altered condition of the animal is beautifully shown in the instance now under consideration, the nature of which will be easily understood by a reference to the following diagrams, giving the exact relative proportions of the different cavities of the stomach to each other in the young calf and in the full-grown cow.

[I am informed by Professor Symonds, of the Royal Veterinary College, that the two following sketches should be placed in the page so as to be viewed with the œsophagus to the right, and the pylorus to the left, instead of being, as they now are, at the top and the bottom; but as the present object is only to show the relative sizes of the different cavities, the error is not of much consequence.]

[Pg 12]

The letters refer to the same parts in each figure: *a*, the paunch; *b*, the honeycomb bag; *c*, the many-plies; *d*, the reed.

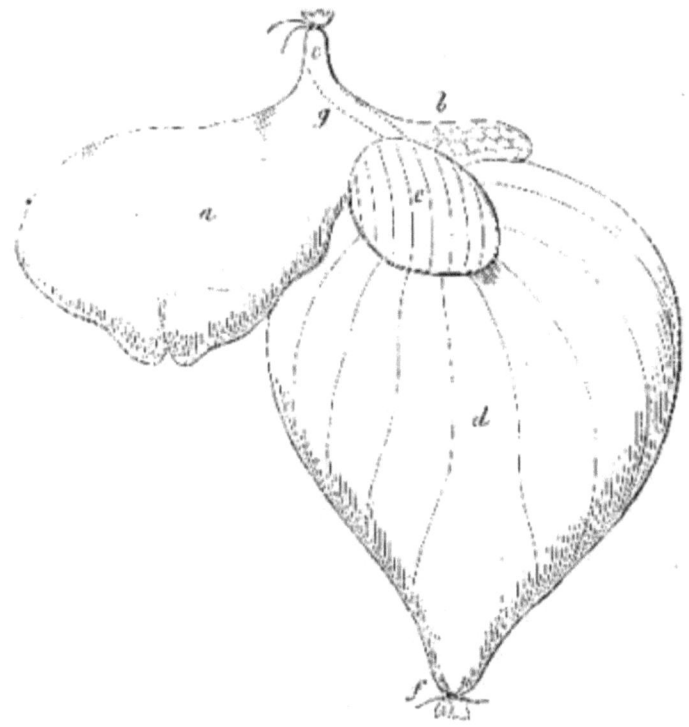

Outline of the Stomach of a Calf about a fortnight old.

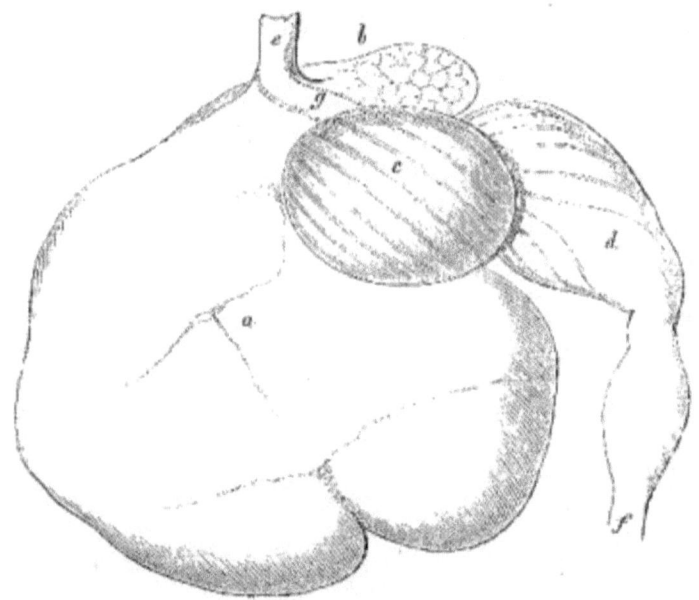

Outline of the Stomach of a full-grown Cow.

[These engravings, illustrative of the comparative sizes of the different stomachal cavities, are copied from original drawings taken from preparations of the stomachs which I made expressly for this purpose.]

In all herbivorous animals, and especially those of the ruminating kind, the alimentary canal is of an enormous length; measuring in a full grown ox, as much as sixty yards. The paunch, in such an animal, will hold from fifteen to eighteen gallons.

Blumenbach observes, that the process of rumination supposes a power of voluntary motion in the œsophagus; and, indeed, the influence of the will throughout the whole process is incontestible. It is not confined to any particular time, since the animal can delay it according to circumstances, even when the paunch is quite full. It has been expressly stated of some men, who have had the power of ruminating, that it was quite voluntary with them. Blumenbach knew four men who ruminated their food, and they assured him they had a real enjoyment in doing it: two of them had the power of doing or abstaining from it at their pleasure.

A case of human rumination occurred some years ago at Bristol, the particulars of which are minutely recorded in the 'Philosophical Transactions.' It seemed, in this instance, to have been hereditary, as the father of the individual was subject to the same habit. The young man usually began to chew his food over again, within a quarter of an hour after eating. His ruminating after a full meal generally lasted about an hour and a half; nor could he sleep until this task was completed. The victuals, upon its return, tasted even more pleasantly than at first; and seemed as if it had been beaten up in a mortar. If he ate a variety of things, that which he ate first, came up again first; and if this return was interrupted for any length of time, it produced sickness and disorder; nor was he ever well till it returned. These singular cases are caused, no doubt, by some abnormal structure of the [Pg 15] interior of the stomach. No account has yet been given of the dissection of an individual so constituted.

When cattle are at rest, or not employed in grazing or chewing the cud, they are observed frequently to lick themselves. By this means they raise up the hair of their coats, and often swallow it in considerable quantities. The hair thus swallowed gradually accumulates in the stomach, where it is formed into smooth round balls, which, in time, become invested with a hardish brown crust, composed, apparently, of inspissated mucilage, that, by continual friction from the coats of the stomach, becomes hard and glossy. It is generally in the paunch that these hair-balls are found. They vary in weight from a few ounces to six or seven pounds. Mr. Walton, author of an 'Account of the Peruvian Sheep,' makes mention of one that he had in his possession which weighed eight pounds and a quarter. This hair-ball had been taken from a cow that fed on the Pampas of Buenos Ayres. It was of a flat circular shape, and measured two feet eleven inches and a half in circumference; two feet eight inches round the flat part; nine inches diameter also in the flat part; eleven inches diameter in the cross part; and, on immersing it in water, it displaced upwards of eight quarts, which made its bulk correspond to 462 cubic inches. The digestive functions are sometimes seriously impaired by these concretions; a loss of appetite ensues, and general debility.

In the Museum of Daniel Crosthwaite, there is a very extraordinary ball of hair, taken from a fatted [Pg 16] calf only seven weeks old. The ball of hair, when taken out of the animal's stomach, and full of moisture, weighed eleven ounces. The calf was fatted by Daniel Thwaite, of Dale Head Hall, within six miles of Keswick; and slaughtered by John Fisher, butcher, Keswick. The calf was a particularly healthy animal.

Before closing this brief sketch of the digestive apparatus of the ox, it may not be uninteresting to quote some of the quaint speculations of Nathaniel Grew on this subject, from his 'Comparative Anatomy of Stomachs and Guts.'

He says: "The *voluntary* motion of the stomach is that only which accompanies rumination. That it is truly voluntary, is clear, from the command that ruminating animals have of that action. For this purpose it is, that the muscles of their venters are so thick and strong; and have several duplicatures, as the bases of those muscles, whereupon the stress of their motion lies. By means whereof they are able with ease to rowl and tumble any part of the meat from one cell of the same venter to another; or from one venter to another; or from thence into the gullet, whensoever they are minded to do it; so that the ejectment of the meat, in rumination, is a voluntary eructation.

"The pointed knots, like little papillæ, in the stomachs of ruminating beasts, are also of great use, namely, for the tasting of the meat. The inner membrane of the first three venters is fibrous (like the gustatory papillæ of the tongue) and not glandulous; the fourth only being glandulous, as in a man. Of [Pg 17] the fibres of this membrane, and the nervous, are composed those pointed knots, which are, both in substance and shape, altogether like to those upon the tongue. Whence I doubt not, but that the said three ventricles, as they have a power of voluntary motion, so, likewise, that they are the seat of taste, and as truly the organs of that sense, as is the tongue itself."

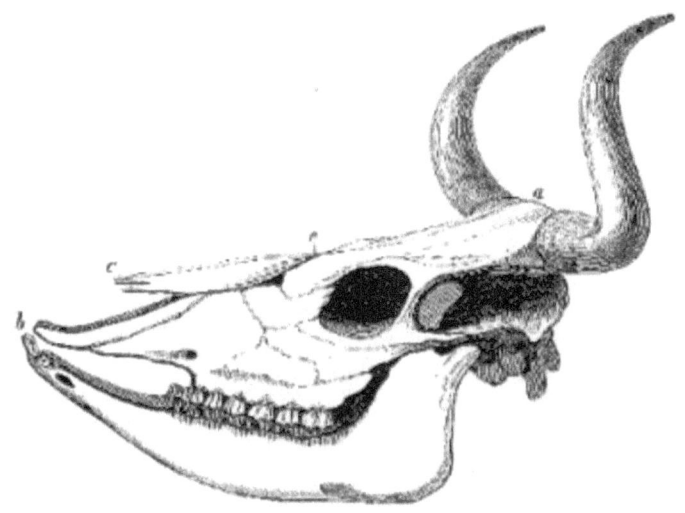

Skull of Domestic Ox, from a specimen in the Royal College of Surgeons.

The mouth of animals of the Ox Tribe contains, when full, thirty-two teeth. Six molars in each jaw, above, below, and on either side; and eight incisors in the [Pg 18] lower jaw. In the upper jaw there are no incisors; but instead thereof a fibrous and elastic pad, or cushion, which covers the convex extremity of the anterior maxillary bone, and which is well worthy of observation.

The final cause of this pad (which stands in the place of upper incisor teeth) and the part it plays in the procuring of food, is thus described by Youatt. "The grass is collected and rolled together by means of the long and moveable tongue; it is firmly held between the lower cutting teeth and the pad, the cartilaginous upper lip assisting in this; and then by a sudden nodding motion of the head, the little roll of herbage is either torn or cut off, or partly both torn and cut.

"The intention of this singular method of gathering the food, it is somewhat difficult satisfactorily to explain. It is peculiar to ruminants, who have one large stomach, in which the food is kept as a kind of reservoir until it is ready for the action of the other stomachs. While it is kept there it is in a state of maceration; it is exposed to the united influence of moisture and warmth, and the consequence of this is, that a species of decomposition sometimes commences, and a vast deal of gas is extricated.

"That this should not take place in the natural process of retention and maceration, nature possibly established this mechanism for the first gathering of the food. It is impossible that half of that which is thus procured can be fairly cut through; part will be torn, and no little portion will be torn up by the roots. [Pg 19] If cattle are observed while they are grazing, it will be seen that many a root mingles with the blades of grass; and these roots have sometimes no inconsiderable quantity of earth about them. The beast, however, seems not to regard this; he eats on, dirt and all, until his paunch is filled.

"It was designed that this earth should be gathered and swallowed; it was the meaning of this mechanism. A portion of absorbent earth is found in every soil, sufficient not only to prevent the evil that would result from occasional decomposition, by neutralizing the acid principle as rapidly as it is evolved; but, perhaps, by its presence, preventing that decomposition from taking place. Hence the eagerness with which stall-fed cattle, who have not the opportunity of plucking up the roots of grass, evince for mould. It is seldom that a cow will pass a newly-raised mole hill without nuzzling into it, and devouring a considerable portion of it. This is particularly the case where there is any degree of indigestion."

The general disposition of animals of this class, when unmolested, is inoffensive and retiring; but when excited and irritated, they are fierce and courageous, and extremely dangerous to encounter. It is a remarkable circumstance in their history, that they are generally provoked to attack at the sight of red, or any very bright and glaring colour. [Pg 20]

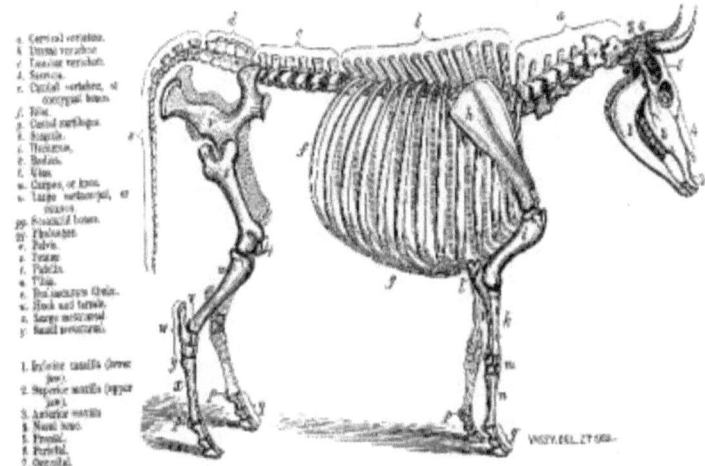

Skeleton of Domestic Ox, from a specimen in the Royal College of Surgeons.

[Pg 21]

THE OX TRIBE

OR

Genus **BOS,**

Is distinguished from other Genera of Ruminantia by possessing hollow persistent horns, growing on a bony core; the tail long, terminated by a tuft of hair; and four inguinal mammæ.

THE AMERICAN BISON.

Bos Americanus.

THE BISON.

The head of this animal is enormously large; larger, in fact, in proportion to the size of its body, than that of [Pg 22] any other species of the Ox Tribe. This huge head is supported by very powerful muscles, attached to the projecting spinous processes of the dorsal vertebræ; and these muscles, together with a quantity of fat, constitute the hump on the shoulders. The horns are short, tapering, round, and very distant from each other, as are also the eyes, which are small and dark. The head, neck, shoulders, and fore-legs, to the knee-joints, are covered with long woolly hair, which likewise forms a beard under the mouth. The rest of the body is clothed only by short, close hair, which becomes rather woolly in the depth of winter. The colour is of a deep brown, nearly black on the head, and lighter about the neck and shoulders. The legs are firm and muscular; the tail is short, with a tuft at the end.

The female is, in every respect, much smaller than the male; her horns are more slender, and the hair on her neck and shoulders is not so thick or long, nor the colour so dark. She brings forth in the spring, and rarely more than one. The calves continue to be suckled nearly twelve months, and follow the cows for a much longer period. It is said that the cows are not unfrequently followed by the calves of two, or even three, breeding seasons.

These animals, both male and female, are timid and shy, notwithstanding their fierce appearance; unless they are wounded, or during the breeding season, when it is dangerous to approach. Their mode of attack is to throw down, by pushing, as they run with their head; then to crush, by trampling their enemy under their fore-feet, which, surmounted as they are, by their tremendous head and shoulder, form most effectual weapons of destruction. [Pg 23]

Young female Bison, after Cuvier.

The following account, by Dr. Richardson, affords an instance of the danger to be apprehended from these powerful animals, when wounded, and not disabled: "Mr. Finnan M'Donald, one of the Hudson's Bay Company's clerks was descending the Saskatchewan in a boat; and one evening, having pitched his tent for the night, he went out in the dusk to look for game. It had become nearly dark when he fired at a Bison bull, which was galloping over an eminence; and as he was hastening forward to see if this shot had taken effect, the wounded beast made a rush at him. He had the presence of mind to seize the animal by the long hair on the forehead, as it struck him on the side with its horn, and being a remarkably tall and powerful man, a struggle ensued, which continued until his wrist was severely sprained, and his arm was rendered powerless; he then fell, and after receiving two or three blows, became senseless. Shortly afterwards he was found by [Pg 24] his companions, lying bathed in his blood, being gored in several places, and the Bison was couched beside him, apparently waiting to renew the attack, had he shown any signs of life. Mr. M'Donald recovered from the immediate effects of the injuries, but he died a few months afterwards. Many instances might be mentioned of the tenaciousness with which this animal pursues its revenge; and I have been told of a hunter being detained for many hours in a tree, by an old bull, which had taken its post below, to watch him."

Wounded Bison, after Catlin.

The capture of the Bison is effected in various ways, chiefly with the rifle, and on foot. Their sense of smelling, however, is so acute, that they are extremely difficult of approach, scenting their enemy from afar, and retiring with the greatest precipitation. Care, therefore, must be taken to go against the wind, in which case they may be approached very near, being almost blinded by the long hair hanging over their foreheads. The hunters generally aim at the shoulder, which, if effectually hit, [Pg 25] causes them to drop at once; otherwise they are infuriated, and become dangerous antagonists, as was proved in the result of Mr. M'Donald's adventure.

When flying before their pursuers, it would be in vain for the foremost to halt, or attempt to obstruct the progress of the main body, as the throng in the rear, still rushing onwards, the leaders must advance, although destruction await the movement. The Indians take advantage of this circumstance to destroy great quantities of this favorite game; and certainly no method could be resorted to more effectually destructive, nor could a more terrible devastation be produced, than that of forcing a numerous herd of these large animals to leap from the brink of a dreadful precipice upon a rocky and broken surface, a hundred feet below.

When the Indians determine to destroy Bisons in this way, one of their swiftest-footed and most active young men is selected, who is disguised in a Bison skin, having the head, ears, and horns adjusted on his own head, so as to make the deception very complete; and thus accoutred, he stations himself between the Bison herd and some of the precipices, which often extend for several miles along the rivers. The Indians surround the herd as nearly as possible, when, at a given signal, they show themselves, and rush forward with loud yells. The animals being alarmed, and seeing no way open but in the direction of the disguised Indian, run towards him, and he, taking to flight, dashes on to the precipice, where he suddenly secures himself in some previously ascertained crevice. The foremost of the herd arrives at the brink,—there is no possibility of retreat, no chance of escape; the foremost may, for an instant, shrink with [Pg 26] terror, but the crowd behind, who are terrified by the approaching hunters, rush forward with increasing impetuosity, and the aggregate force hurls them successively into the gulf, where certain death awaits them.

Sometimes they are taken by the following method:—A great number of men divide and form a vast square; each band then sets fire to the dry grass of the savannah, where the herds are feeding; seeing the fire advance on all sides, they retire in great consternation to the centre of the square; the men then close and kill them without the least hazard.

Great numbers are also taken in pounds, constructed with an embankment of such an elevation as to prevent the return of the Bisons when once they are driven into it. A general slaughter then takes place with rifles or arrows.

The following vivid sketch is from the narrative of John Tanner, who, when about seven or eight years of age, was stolen from his parents by the Indians, and remained with them during a period of thirty years.

"By the end of the second day after we left Pembinah we had not a mouthful to eat, and were beginning to be very hungry. When we laid down in our camp (near Craneberry River) at night, and put our ears close to the ground, we could hear the tramp of the buffaloes, but when we sat up we could hear nothing; and on the following morning nothing could be seen of them; though we could command a very extensive view of the prairie. As we knew they must not be far off in the direction of the sounds we had heard, eight men, of whom I was one, were selected and dispatched to kill some, and bring the meat to a point where it was agreed the [Pg 27] party should stop next night. The noise we could still hear next morning, by applying our ears to the ground; and it seemed about as far distant, and in the same direction, as before. We started early, and rode some hours before we could begin to see them; and when we first discovered the margin of the herd, it must have been at least ten miles distant. It was like a black line drawn along the edge of the sky, or a low shore seen across a lake. The distance of the herd from the place where we first heard them could not have been less than twenty miles. But it was now the rutting season, and various parts of the herd were all the time kept in rapid motion by the severe fights of the bulls. To the noise produced by the knocking together of the two divisions of the hoof, when they raised their feet from the ground, and of their incessant tramping, was added the loud and furious roar of the bulls, engaged, as they all were, in their terrific and appalling conflicts. We were conscious that our approach to the herd would not occasion the alarm now, that it would at any other time, and we rode directly towards them. As we came near we killed a wounded bull, which scarcely made an effort to escape from us. He had wounds in his flanks, into which I could put my whole hand. As we knew that the flesh of the bulls was not now good to eat, we did not wish to kill them, though we might easily have shot any number. Dismounting, we put our horses in the care of some of our number, who were willing to stay back for that purpose, and then crept into the herd to try to kill some cows. I had separated from the others, and advancing, got entangled among the bulls. Before I found an opportunity to shoot a cow, the bulls began [Pg 28] to fight very near me. In their fury they were totally unconscious of my presence, and came rushing towards me with such violence, that in some alarm for my safety, I took refuge in one of those holes which are so frequent where those animals abound, and which they themselves dig to wallow in. Here I found they were pressing directly upon me, and I was compelled to fire to disperse them, in which I did not succeed until I had killed four of them. By this firing the cows were so frightened, that I perceived I should not be able to kill any in this quarter; so regaining my horse, I rode to a distant part of the herd, where the Indians had succeeded in killing a fat cow. But from this cow, as is usual in similar cases, the herd had all moved off, except one bull, who, when I came up, still kept the Indians at bay. 'You are warriors,' said I, as I rode up, 'going far from your own country, to seek an enemy, but you cannot take his wife from that old bull, who has nothing in his hands.' So saying, I passed them directly towards the bull, then standing something more than two hundred yards distant. He no sooner saw me approach, than he came plunging towards me with such impetuosity, that, knowing the danger to my horse and myself, I turned and fled. The Indians laughed heartily at my repulse, but they did not give over their attempts to get at the cow. By dividing the attention of the bull, and creeping up to him on different sides, they at length shot him down. While we were cutting up the cow, the herd were at no great distance; and an old cow, which the Indians supposed to be the mother of the one we had killed, taking the scent of the blood, came running with great violence towards us. The Indians were alarmed and fled, many of them not having their [Pg 29] guns in their hands; but I had carefully reloaded mine, and had it ready for use. Throwing myself down close to the body of the cow, and behind it, I waited till the other came up within a few yards of the carcase, when I fired upon her; she turned, gave one or two jumps, and fell dead. We had now the meat of two fat cows, which was as much as we wanted; accordingly we repaired, without delay, to the appointed place, where we found our party, whose hunger was already somewhat allayed by a deer one of them had killed."

In hunting the Bison, the spear and the arrow are still much in use among the Indians. The following sketch (after Catlin) represents an Indian in the act of shooting a Bison with the arrow: —

In the 'Letters and Notes on the North-American Indians,' by Catlin, there are a great many interesting details of the Bison (or Buffalo, as it is there called).

"Six days of severe travelling have brought us from the Camanchee village to the north bank of the Canadian, where we are snugly encamped on a beautiful plain, and in the midst of countless numbers of buffaloes; and halting a few days to recruit our horses and men, and dry meat to last us the remainder of our journey. [Pg 30]

"The plains around this, for many miles, seem actually speckled, in distance and in every direction, with herds of grazing buffaloes; and for several days, the officers and men have been indulged in a general license to gratify their sporting propensities; and a scene of bustle and cruel slaughter it has been, to be sure! From morning till night, the camp has been daily almost deserted. The men have dispersed in little squads, in all directions, and are dealing death to these poor creatures to a most cruel and wanton extent, merely for the pleasure of destroying, generally without stopping to cut out the meat. During yesterday and to day, several hundreds have undoubtedly been killed, and not so much as the flesh of half a dozen used. Such immense swarms of them are spread over this tract of country, and so divided and terrified have they become, finding their enemies in all directions where they run, that the poor beasts seem completely bewildered, running here and there, and, as often as otherwise, come singly advancing to the horsemen, as if to join them for their company, and are easily shot down. In the turmoil and confusion, when their assailants have been pushing them forward, they have galloped through our encampment, jumping over our fires, upsetting pots and kettles, driving horses from their fastenings, and throwing the whole encampment into the greatest consternation and alarm."

Speaking of the attacks made upon them by the Wolves, he says, "When the herd is together the Wolves never attack them, as they instantly gather for combined resistance, which they effectually make. But when the herds are travelling, it often happens that an aged or wounded one lingers at a little distance behind, and when [Pg 31] fairly out of sight of the herd, is set upon by the voracious hunters, which often gather to the number of fifty or more, and are sure at last to torture him to death, and use him up at a meal. The Buffalo, however, is a huge and furious animal, and when his retreat is cut off, makes desperate and deadly resistance, contending to the last moment for the right of life, and oftentimes deals death by wholesale to his canine assailants.

"During my travels in these regions, I have several times come across such a gang of these animals surrounding an old or wounded bull, where it would seem, from appearances, that they had been for several days in attendance, and at intervals desperately engaged in the effort to take his life. But a short time since, as one of my hunting companions and myself were returning to our encampment, with our horses loaded with meat, we discovered at a distance a huge bull, encircled with a gang of white wolves. We rode up as near as we could without driving them away; and being within pistol-shot, we had a remarkably good view, where I sat for a few moments and made a sketch in my note-book. After which we rode up, and gave the signal for them to disperse, which they instantly did, withdrawing themselves to the distance of fifty or sixty rods, when we found, to our great surprise, that the animal had made desperate resistance, until his eyes were entirely eaten out of his head; the gristle of his nose was mostly gone; his tongue was half eaten off, and the skin and flesh of his legs torn almost literally into strings. In this tattered and torn condition the poor old veteran stood bracing up in the midst of his devourers, who had ceased hostilities for a few minutes, to enjoy a sort of parley, [Pg 32] recovering strength to resume the attack in a few moments again. In this group, some were reclining to gain breath, whilst others were sneaking about, and licking their chaps in anxiety for a renewal of the attack; and others, less lucky, had been crushed to death by the feet or the horns of the bull. I rode nearer to the pitiable object, as he stood bleeding and trembling before me, and said to him,—"Now is your time, old fellow, and you had better be off." Though blind, and nearly destroyed, he straightened up, and, trembling with excitement, dashed off at full speed upon the prairie, in a straight line. We turned our horses, and resumed our march; and when we had advanced a mile or more, we looked back, and again saw the ill-fated animal surrounded by his tormentors, to whose insatiable voracity he unquestionably soon fell a victim."

Bison surrounded by Wolves, after Catlin.

It has frequently been noticed, that whenever a female Bison, having a calf, is slain, the young one remains by its fallen dam, with signs of strong natural affection, [Pg 33] and instinctively follows the inanimate carcase of its parent to the residence of the hunter. In this way many calves are secured.

According to Mr. Catlin's account these young animals are induced to follow any one who merely breathes in their nostrils. "I have often," says he, "in concurrence with a known custom of the country, held my hands over the eyes of the calf, and breathed a few strong breaths into its nostrils; after which I have, with my hunting companions, rode several miles into our encampment, with the little prisoner busily following the heels of my horse the whole way, as closely as its instinct would attach it to the company of its dam.

Bison Calf, about three weeks old.

"This is one of the most extraordinary things that I have met with in the habits of this wild country; and although I had often heard of it, and felt unable exactly to believe it, I am now willing to bear testimony to the fact, from the numerous instances which I have witnessed [Pg 34] since I came into the country. During the time that I resided at this post (Teton River) in the spring of the year, on my way up the river, I assisted in bringing in, in the above manner, several of these little prisoners, which sometimes followed for five or six miles close to our horse's heels, and even into the Fur Company's Fort, and into the stable where our horses were led. In this way, before I left for the head waters of the Missouri, I think we had collected about a dozen, which Mr. Laidlaw was successfully raising with the aid of a good milch cow, and which were to be committed to the care of Mr. Chouteau, to be transported, by the return of the steamer, to his extensive plantation in the vicinity of St. Louis."

The uses which are made of the various parts of the Bison are numerous. The hide, which is thick and rather porous, is converted by the Indians into mocassins for the winter; they also make their shields of it. When dressed with the hair on, it is made into clothing by the natives, and most excellent blankets by the European settlers; so valuable, indeed, is it esteemed, that three or four pounds sterling a piece are not unfrequently given for good ones in Canada, where they are used as travelling cloaks. The fleece, which sometimes weighs eight pounds, is spun and wove into cloth. Stockings, gloves, garters, &c., are likewise knit with it, appearing and lasting as well as those made of the best sheep's wool. In England it has been made into remarkably fine cloth.

"There are," says Catlin, "by a fair calculation, more than 300,000 Indians who are now subsisting on the flesh of the buffaloes, and by these animals supplied [Pg 35] with, all the luxuries of life which they desire, as they know of none others. The great variety of uses to which they convert the body and other parts of that animal, are almost incredible to the person who has not actually dwelt amongst these people, and closely studied their modes and customs. Every part of their flesh is converted into food, in one shape or other, and on it they entirely subsist. The skins of the animals are worn by the Indians instead of blankets; their skins, when tanned, are used as coverings for their lodges and for their beds; undressed, they are used for constructing canoes, for saddles, for bridles, l'arrêts, lasos, and thongs. The horns are shaped into ladles and spoons; the brains are used for dressing the skins; their bones are used for saddle-trees, for war-clubs, and scrapers for graining the robes; and others are broken up for the marrow fat which is contained in them. The sinews are used for strings and backs to their bows, for thread to string their beads and sew their dresses. The feet of the animals are boiled, with their hoofs, for the glue they contain, for fastening their arrow points, and many other uses. The hair from the head and shoulders, which is long, is twisted and braided into halters, and the tail is used for a fly-brush."

Again (vol. ii, p. 138), he says, "I have introduced the skin canoes of the Mandans (of the Upper Missouri), which are made almost round like a tub, by straining a buffalo's skin over a frame of wicker-work, made of willow or other boughs. The woman, in paddling these awkward tubs, stands in the bow, and makes the stroke with the paddle, by reaching it forward in the water, and drawing it to her, by which means she pulls the [Pg 36] canoe along with considerable speed. These very curious and rudely-constructed canoes are made in the form of the Welsh coracle; and, if I mistake not, propelled in the same manner, which is a very curious circumstance; inasmuch as they are found in the heart of the great wilderness of

America, where all the surrounding tribes construct their canoes in decidedly different forms, and of different materials."

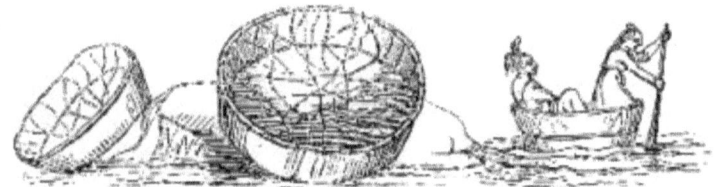

Skin Canoes of the Mandan Indians.

It is generally agreed by travellers, that the flesh of the Bison is little inferior to the beef of our domestic oxen. The tongue is considered a delicacy, and the hump is much esteemed. A kind of potted-beef, called *pemmican*, is made of the flesh of the Bison, in the following manner: — The flesh is spread on a skin, dried in the sun, and pounded with stones; then all the hair is carefully sifted out of it, and melted fat kneeded into it. This, when properly made and kept dry, will keep good for twelve months. The tallow of the Bison forms an important article of commerce; one fat bull yielding sometimes as much as 150 pounds weight.

Mr. Turner, a gentleman long resident in America, is of opinion, that the Bison is superior even to our domestic cattle for the purposes of husbandry, and has expressed a wish to see this animal domesticated on the [Pg 37] English farms. He informs us, that a farmer on the great Kenhawa broke a young Bison to the plough; and having yoked it with a steer, taken from his tame cattle, it performed its work to admiration. But there is another property in which the Bison far surpasses the Ox, and this is his strength. "Judging from the extraordinary size of his bones, and the depth and formation of the chest, (continues this gentleman,) I should not think it unreasonable to assign nearly a double portion of strength to this powerful inhabitant of the forest. Reclaim him, and you gain a capital quadruped, both for the draught and for the plough; his activity peculiarly fits him for the latter, in preference to the ox."

As there are no Game Laws in America, (except in a very few confined instances on the Atlantic border,) the consequence is that the Bison is fast disappearing before the approach of the white settlers.

At the commencement of the eighteenth century these wild cattle were found in large numbers all throughout the valley of the Ohio, of the Mississippi, in Western New York, in Virginia, &c. In the beginning of the present century they were still existing in the extreme western or southwestern part of the State of New York. As late as 1812 they were natives of Ohio, and numerous in that State. And now they are not to be seen in their native state in any part of the United States, east of the Mississippi River; nor are they now to be found in any considerable numbers west of that great river, until you have travelled some eighty or a hundred miles into the interior of the country.

There were no Bisons west of the Rocky Mountains, when Lewis and Clarke travelled there in 1805. On [Pg 38] their return from the Columbia, or Oregon River, in July of that year, the first Bison they saw was on the day after they commenced their descent of the Rocky Mountains towards the east. On the second day after that, they saw immense herds of them on the banks of the Medicine River. One collection of these animals which they subsequently saw, on the borders of the Missouri River, they estimated as being at least 20,000 in number.

In 1823 it was discovered that the Bisons had crossed the Rocky Mountains, and some were to be seen in the vallies to the west of that range.

East of that range of mountains, these animals migrate from the uplands or mountains to the plains, and from north to south, about the beginning of November; and return from the south to the north, and from the plains to the uplands, soon after the disappearance of the snow in the spring.

The herds of Bisons wander over the country in search of food, usually led by a bull remarkable for strength and fierceness. While feeding, they are often scattered over a great extent of country; but when they move, they form a dense and almost impenetrable column, which, when once in motion, is scarcely to be impeded. Their line of march is seldom interrupted, even by considerable rivers, across which they swim, without fear or hesitation, nearly in the order in which they traverse the plains. The Bisons which frequent the woody parts of the country form smaller herds than those which roam over the plains, but are said to be individually of a greater size.

The rutting takes place the latter part of July and [Pg 39] the beginning of August, after which the cows separate from the bulls in distinct herds. They bring forth their young in April: from which it appears that the term of gestation is about nine months.

The pair of American Bisons in the Zoological Gardens produced a calf in 1849; from the observations made in that instance, the period of gestation was calculated at 270 days.

The most important anatomical difference between the American and the European is, that the American has fifteen pairs of ribs, whereas the European has but fourteen.

The following are the dimensions of a large specimen: —

	Ft.	In.
From the nose to the insertion of the tail	8	6
Height at the shoulder	6	0
" at the croup	5	0
Length of the head	2	1

Their weights vary from 1200 to 2000 pounds.

Head of young male Bison.

THE AUROCHS, OR EUROPEAN BISON.

Bos Bison.

In this, as in the American species, the head is very broad, and the forehead arched; but the horns are longer, more curved, and end in a finer point than those of the American Bison. The eyes are large and dark; the hair on the forehead is long and wavy; under the chin and on the breast it forms a sort of beard. In winter, the whole of the neck, hump, and shoulders are covered with a long woolly hair of a dusky brown colour, intermingled with a short soft fur of a fawn colour. The long hair is gradually cast in the summer, to be again renewed as the inclemency of winter comes on. The legs, back, and [Pg 41] posterior portions are covered with short, dark brown hair. The tail is of a moderate length, is covered with hair, and terminates in a large tuft.

The females are not so large as the males, neither are they characterised by that abundance of hair on the anterior parts, which is so conspicuous in the bulls.

These animals have never been domesticated, although calves have sometimes been caught, and confined in an enclosed pasture. An instance of this kind is recorded by Mr. Gilibert, who, while in Poland, had the opportunity of observing the character of four young ones thus reared in captivity. They were suckled by a she-goat, obstinately refusing to touch a common cow. This antipathy to the domestic cow, which they manifested so early, maintained its strength as they advanced in years; their anger was sure to be excited at the appearance of any domestic cattle, which, whenever introduced to them, they vigorously expelled from their pasture. They were, however, sufficiently tame to acknowledge the voice of their keeper.

The geographical range of this animal is now comparatively very limited, being confined to the forests of Lithuania, Moldavia, Wallachia, and some of the Caucasian mountain forests; yet there can be no doubt that, at an early period, they roamed at large over a great part of both Europe and Asia.

Although they have never been, strictly speaking, domesticated, yet herds of them are kept in certain localities in the forest of Bialowieza, under the special protection of the Emperor of Russia, and under the immediate superintendence of twelve herdsmen, each herdsman keeping the number allotted to his charge in a particular department of the forest, near some river [Pg 42] or stream. The estimated number of the twelve herds is about 800.

They feed on grass and brushwood; also on the leaves and bark of young trees, particularly the willow, poplar, ash, and birch. In autumn they likewise browse on heath, and the lichens which cover the bark of trees. In winter, when the ground is covered with snow, fodder is provided for them.

Their cry is quite peculiar, resembling a groan, or a grunt, more than the lowing of an ox.

They do not attain their full stature until after the sixth year, and live till between thirty and forty.

"The strength of the Zubr," says Dr. Weissenborn, "is enormous; and trees of five or six inches diameter cannot withstand the thrusts of old bulls. It is neither afraid of wolf nor bear, and assails its enemies both with its horns and hoofs. An old Zubr is a match for four wolves; packs of the latter animal, however, sometimes hunt down even old bulls when alone; but a herd of Zubrs has nothing to fear from any rapacious animal.

"Notwithstanding the great bulk of its body, the Zubr can run very swiftly. In galloping, its hoofs are raised above its head, which it carries very low. The animal has, however, but little bottom, and seldom runs farther than one or two English miles. It swims well, and is very fond of bathing.

"The zubr is generally exceedingly shy, and avoids the approach of man. They can only be approached from the leeward, as their smell is extremely acute. But when accidentally and suddenly fallen in with, they will passionately assail the intruder. In such fits of passion the animal thrusts out its tongue repeatedly, lashes its [Pg 43] sides with its tail, and the reddened and sparkling eyes project from their sockets, and roll furiously. Such is their innate wildness, that none of them have been completely tamed. When taken young they become, it is true, accustomed to their keepers, but the approach of other persons renders them furious; and even their keepers must be careful always to wear the same sort of dress when going near them. Their great antipathy to the Bos Taurus, which they either avoid or kill, would render their domestication, if it were practicable, but little desirable. The experiments made with a view of obtaining a mixed breed from the Zubr and Bos Taurus have all failed, and are now strictly prohibited."

The rutting season is in August, and continues for about a fortnight; the calves are produced in May; thus, the period of gestation is between nine and ten months. The calves continue to suckle nearly twelve months, and the cows seldom calve oftener than once in three years.

The European Bison differs internally from the common ox in having fourteen pairs of ribs, whereas the common ox has but thirteen. The external differences between the two animals are too obvious to require pointing out.

In 1845, the Emperor of Russia presented to the British Museum a very fine stuffed specimen of this animal, from which the figure at the head of this chapter was taken.

The following are its dimensions: —

	Ft.	In.
Length from the nose to the insertion of the tail	9	10
Height at the withers	5	6

"	at the rump	4	11
Length of head		1	8
"	of tail	3	0

[Pg 44]

M. Dimitri de Dolmatoff, Master of the Imperial Forests in the Government of Grodno, in his note of the capture of the Aurochs, (written in 1847,) alludes to the statement (made by every writer who has treated of these animals), that the calves, although taken young, invariably refuse to be suckled by the Domestic Cow. This he contradicts in the most explicit manner, on the testimony of his own experience, having had several instances come under his observation, in which the young calves of the Aurochs were suckled and reared by cows of the common domestic species.

Cæsar, in his account of the "Sylva Hercynia" — the Black Forest — thus mentions the Urus, amongst other animals, there found:

"A third kind [of animals] are those called Uri. They are but little less than Elephants in size, and are of the species, colour, and form of a bull. Their strength is very great, and also their speed. They spare neither man nor beast that they see. They cannot be brought to endure the sight of men, nor be tamed, even when taken young. The people who take them in pit-falls, assiduously destroy them; and young men harden themselves in this labour, and exercise themselves in this kind of chase; and those who have killed a great number — the horns being publicly exhibited in evidence of the fact — obtain great honour. The horns, in amplitude, shape, and species, differ much from the horns of our oxen. They are much sought after; and after having been edged with silver at their mouths, they are used for drinking vessels at great feasts." (*De Bello Gallico*, lib. vi.)

[Pg 45]

THE YAK, OR SOORA-GOY.

Bos Grunniens.

The following interesting and circumstantial account of this curious species of Ox, is from the pen of Lieut. Samuel Turner. (*Asiatic Researches*, vol. iv.)

"The Yak of Tartary, called Soora-Goy in Hindostan, and which I term the Bushy-tailed Bull of Tibet, is about the height of an English Bull, which he resembles in the figure of the body, head, and legs. I could distinguish between them no essential difference, except only that the Yak is covered all over with a thick coat of long hair. The head is rather short, crowned with two smooth round horns, that, tapering from the setting on, terminate in sharp points, arch inwardly, and near the extremities are a little turned back. The ears are small; the forehead appears prominent, being adorned with much curling hair; the eyes are full and large; the nose smooth and convex; [Pg 46] the nostrils small. The neck is short, describing a curvature nearly equal both above and below; the withers high and arched; the rump low. Over the shoulders rises a bunch, which at first sight would seem to be the same kind of exuberance peculiar to the cattle of Hindostan; but in reality it consists in the superior length of the hair only, which, as well as that along the ridge of the back to the setting on of the tail, grows long and erect, but not harsh. The tail is composed of a prodigious quantity of long flowing glossy hair, descending to the hock; and is so extremely well furnished, that not a joint of it is perceptible; but it has much the appearance of a large bunch of hair artificially set on. The shoulders, rump, and upper part of the body are clothed with a sort of thick soft wool, but the inferior parts with straight pendent hair that descends below the knee; and I have seen it so long in some cattle, which were in high health and condition, as to trail along the ground. From the chest, between the fore-legs, issues a large pointed tuft of hair, growing somewhat larger than the rest. The legs are very short. In every other respect, hoofs, &c., he resembles the ordinary Bull. There is a great variety of colours among them, but black and white are the most prevalent. It is not uncommon to see the long hair upon the ridge of the back, the tail, the tuft upon the chest, and the legs below the knee white, when all the rest of the animal is jet black.

"These cattle, though not large boned, from the profuse quantity of hair with which they are provided, appear of great bulk. They have a down heavy look, but are fierce, and discover much impatience at the near approach of strangers. They do not low loud (like the [Pg 47] cattle of England) any more than those of Hindostan; but make a low grunting noise, scarcely audible, and that but seldom, when under some impression of uneasiness. These cattle are pastured in the coldest part of Tibet, upon short herbage, peculiar to the tops of mountains and bleak plains. That chain of lofty mountains situated between lat. 27° and 28°, which divides Tibet from Bootan, and whose summits are most commonly covered with snow, is their favourite haunt. In this vicinity the Southern glens afford them food and shelter during the severity of the winter; in milder seasons the Northern aspect is more congenial to their nature, and admits a wider range. They are a very valuable property to the tribes of illiterate Tartars, who live in tents, and tend them from place to place, affording their herdsmen a mode of conveyance, a good covering, and subsistence. They are never employed in agriculture, but are extremely useful as beasts of burden; for they are strong, sure-footed, and carry a great weight. Tents and ropes are manufactured of their hair, and I have seen, though amongst the humblest ranks of herdsmen, caps and jackets worn of their skins. Their tails are esteemed throughout the East, as far as luxury or parade have any influence on the manners of the people; and on the continent of India are found, under the denomination of Chowries, in the hands of the meanest grooms, as well as, occasionally, in those of the first ministers of state. Yet the best requital with which the care of their keepers is at length rewarded for selecting them good pastures, is in the abundant quantity of rich milk they give, yielding most excellent butter, which they have a custom of depositing in skins or bladders, and excluding the air; it keeps in this cold [Pg 48] climate all the year, so that after some time tending their flocks, when a sufficient stock is accumulated, it remains only to load their cattle, and drive them to a proper market with their own produce, which constitutes, to the utmost verge of Tartary, a most material article of commerce."

The soft fur upon the hump and shoulders is manufactured by the natives of Tibet into a fine but strong cloth; and, if submitted to the test of European skill, might no doubt be made to produce a very superior fabric.

The herdsmen commonly convert the hides into a loose outer garment that covers the whole of their bodies, hanging down to the knees; and it proves a sufficient protection against the lowest temperature of the cold and desolate region which they inhabit. It furnishes at once a cloak by day and a bed by night.

The Yak is not generally fierce, but, if intruded upon by strangers, it sometimes manifests very formidable symptoms of impatience, stamping its feet, whisking its tail aloft, and tossing its head. When excited, it is not easily appeased, and is exceedingly tenacious of injury, always showing great fierceness whenever any one approaches who has chanced to provoke it.

The cow is called *Dhe*, of which the wandering Tartars possess great numbers, having no means of subsistence but those supplied by their flocks and herds.

A fine male specimen of this Ox was brought to England by Warren Hastings, and several attempts were made to procure a cross between it and the common English Cow, but without success. He invariably refused to associate with ordinary cattle, and exhibited a decided antipathy to them. His portrait was painted, and is now [Pg 49] in the Museum of the College of Surgeons, London. The following figure (taken from the 'Oriental Annual') is so much like the portrait of Warren Hastings's Yak, that it might almost be taken for a copy of it.

There is the skin of a Yak in the Zoological Museum, which coincides pretty nearly with the foregoing description. There is also a stuffed specimen of a female in the British Museum.

Like the European Bison, the skeleton of the Yak has fourteen pairs of ribs. Period of gestation not recorded.

[Pg 50]

THE GYALL, (*Bos Frontalis* of Lambert;)

THE GAYAL, (*Bos Gavæus* of Colebrooke;)

THE JUNGLY GAU, (*Bos Sylhetanus* of F. Cuvier.)

Of the animals named in the foregoing list, we have had several very interesting accounts; but none of these have been sufficiently precise to enable us to determine the specific character of the animals described.

Are they, as some affirm, merely different names for the same animal; or do they designate animals which are really and truly distinct?

Nothing short of an appeal to structure can satisfactorily settle this or any other disputed point of a similar nature; but, unfortunately for zoology, the opportunities for such appeals are rare, and, when they do occur, are seldom taken advantage of. Let us hope that this hint will not be lost on some of our intelligent countrymen in the East; and that before long we may be favoured with the result of their researches.

In the meantime, and in order to facilitate as much as possible the endeavours of those who may have opportunities for such inquiries, the following epitome is given of the various papers which have already appeared on the subject, but which, in their present scattered form, are of very little general utility.

[Pg 51]

THE GYALL.

The earliest descriptive notice we have of the Gyall was that given in a paper read before the Linnean Society, in 1802, by Mr. Lambert, on the occasion of a bull of this species arriving in London from India.

"Bos Frontalis.

"General colour a blueish-black; the frontal fascia gray; the horns short, thick, and distant at their bases, the tail nearly naked, slender, and with a tuft at the end. The Gyall has no mane; its coat is soft; the edge of the under lip is white, and is fringed with bristling hair. The horns are pale, with their bases included in the frontal fascia."

The Gyall, reduced—from the Linnean Transactions.

The animal of which this description is given, appeared to be between two and three years old, very tame, and inoffensive. [Pg 52] A drawing was taken of it, which was engraved and published in the Linnean Transactions.

The following are its dimensions:

	Ft.	In.
From tip of nose to end of tail	9	2
" tip of hoof of fore foot to top of the rising of back	4	1-1/2
Girth of largest part of abdomen	5	7
From the tip of the hoof of the hind leg to the highest part of the rump	4	0-1/2
" the tip of forehead to end of nose	1	9
Girth of head over the angle of the jaws	2	11-1/2
Between tips of horns	1	8-1/2
Length of horn, externally	0	8-1/2

Girth of horn at largest part   1   1

In reply to some inquiries respecting this animal which he made of a gentleman, (Mr. Harris,) resident in India, Mr. Lambert received the following:

"Dear Sir,—I have before me your note, with the drawing, which undoubtedly appears to me to be the figure of the animal I mentioned to have in my possession. Some parts of the drawing seem to be rather too much enlarged, as in the base of the horns, and the rising between the fore shoulders.

"The animal I described to you, and which I have kept and reared these last seven years, and know by the name of the Gyall, is a native of the hills to the north east and east of the Company's province of Chittagong, in Bengal, inhabiting that range of hills which separates it from the country of Arracan.

"The male Gyall is like our Bull in shape and appearance, but I conceive not quite so tall; it is of a blackish-brown colour; the horns short, but thick and strong towards the base, round which, and across the frons, the [Pg 53] hair is bushy, and of a dirty white colour; the chest and forehead are broad and thick. He is naturally very bold, and will defend himself against any of the beasts of prey.

"The female differs a little in appearance; her horns are not quite so large, and her make is somewhat more slender. She is very quiet, and is used for all the purposes of the dairy; as also, (I have been informed by the natives,) for tilling the ground, and is more tractable than the Buffalo. The milk which these cows give has a peculiar richness in it, arising, I should conceive, from their always feeding on the young shoots and branches of trees in preference to grass.

(Head of Gyall, from Linnean Transactions.)

"I constantly made it a practice to allow them to range abroad, amongst the hills and jungles at Chittagong, during the day, to browse; a keeper attending to prevent their straying so far as to endanger losing them. They do not thrive so well in any part of Bengal as in the [Pg 54] afore-mentioned province, and in the adjoining one, Pipperah, where, I believe, the animal is also to be found. I have heard of a female Gyall breeding with a common Bull. I wish it were in my power to give you more particulars, but I am describing entirely from memory."

In February, 1804, Mr. Lambert again addressed the Linnean Society on the same subject. He says, "Since I presented to the Society the last account of the Bos Frontalis, or Gyall of India, Mr. Fleming, a gentleman who has just returned from that country, has very obligingly communicated to me the following further particulars. This account was transmitted to Mr. Fleming by Mr. Macrae, resident at Chittagong, in a letter, dated March 22, 1802, and was accompanied with a drawing, by which it appears that the animal from which my figure was taken was full grown." (See the figure, p. 51.)

Mr. Macrae's Account.

The Gyall is a species of cow peculiar to the mountains, which form the eastern boundary of the province of Chittagong, where it is found running wild in the woods; and it is also reared as a domestic animal by the Kookies, or Lunclas, the inhabitants of those hills. It delights to live in the deepest jungles, feeding on the tender leaves and shoots of the brushwood; and is never met with on the plains below, except when brought there. Such of them as have been kept by the gentlemen at Chittagong, have always preferred browsing among the thickets on the adjacent hills to feeding on the grass of the plains.

It is of a dull heavy appearance, yet of a form that indicates both strength and activity; and approaches nearly to that of the wild Buffalo. Its head is set on [Pg 55] like the Buffalo's, and it carries it much in the same manner, with the nose projecting forward; but in the shape of the head it differs materially from both the Buffalo and the Cow, the head of the Gyall being much shorter from the crown to the nose, but much broader between the horns than that of either. The withers and shoulders of the Gyall rise higher in proportion than those of Buffalo or Cow, and its tail is small and short, seldom falling lower than the bend in the ham. Its colour is in general brown, varying from a light to a deep shade; it has at times a white forehead, and *white legs*, with a white belly and brush. The hair of the belly is invariably of a lighter colour than that of the back and flanks. The Gyall calf is of a dull red colour, which gradually changes to a brown as it advances in age.

The female Gyall receives the bull at three years of age; her term of gestation is eleven months, when she brings forth, and does not again admit the male until the second year thereafter, thus producing a calf once in three years only. So long an interval between each birth must tend to make the species rare. In the length of time she goes with young, as well as in that between each conception, the Gyall differs from the Buffalo and Cow. The Gyall does not give much milk, but what she yields is nearly as rich as the cream of other milk. The calf sucks its dam for eight or nine months, when it is capable of supporting itself. The Kookies tie up the calf until he is sufficiently strong to do so.

The Gyalls live to the age of from fifteen to twenty. They lose their sight as they grow old, and are subject to a disease of the hoof, which often proves fatal at an early age. When the Kookies consider the disease beyond the [Pg 56] hope of cure, he kills the animal and eats the flesh, which constitutes his first article of luxury.

The Kookies have a very simple method of catching the wild Gyalls, which is as follows: — On discovering a herd of wild Gyalls in the jungles, they prepare a number of balls, of the size of a man's head, composed of a particular kind of earth, salt, and cotton. They then drive their tame Gyalls towards the wild ones, when the two herds soon meet, and assimilate into one; the males of the one attaching themselves to the females of the other, and *vice versâ*. The Kookies now scatter their balls over such parts of the jungle as they think the herd most likely to pass, and watch its motions. The Gyalls, on meeting these balls as they pass along, are attracted by their appearance and smell, and begin to lick them with their tongues; and relishing the taste of the salt, and the particular earth composing them, they never quit the place until all the balls are consumed. The Kookies having observed the Gyalls to have once tasted their balls, prepare a sufficient supply of them to answer the intended purpose; and as the Gyalls lick them up, they throw down more; and it is to prevent their being so readily destroyed that the cotton is mixed with the earth and the salt. This process generally goes on for three changes of the moon, or for a month and a half, during which time the tame and the wild Gyalls are always together, licking the decoy balls; and the Kookie, after the first day or two of their being so, makes his appearance, at such a distance as not to alarm the wild ones. By degrees he approaches nearer and nearer, until at length the sight of him has become so familiar that he can advance to stroke his tame Gyalls on the back and [Pg 57] neck, without frightening away the wild ones. He next extends his hand to them, and caresses them also, at the same time giving them plenty of his decoy balls to lick. Thus, in the short space of time mentioned, he is able to drive them, along with the tame ones, to his parrah, or village, without the least exertion of force; and so attached do the Gyalls become to the parrah, that when the Kookies migrate from one place to another, they always find it necessary to set fire to the huts they are about to abandon, lest the Gyalls should return to them from the new grounds.

It is worthy of remark that the new and full moon are the periods at which the Kookies in general commence their operations of catching the wild Gyalls, from having observed that at these changes the two sexes are most inclined to associate. The same observation has been made with respect to Elephants.

THE GAYAL.

About four years after the publication of Mr. Macrae's account of the Gyall (namely in 1808,) there appeared, in the Eighth volume of 'Asiatic Researches,' a description of a species of Ox, named Gayal, communicated by H. T. Colebrooke.

He commences by observing, that "the Gayal was mentioned in an early volume of the 'Researches of the Asiatic Society,' (vol. ii, p. 188, 1790,) by its Indian name, which was explained by the phrase "Cattle of the mountains." It had been obscurely noticed (if indeed the same species of Ox be meant) by Knox, in his historical relation of Ceylon (p. 21), and it has been imperfectly [Pg 58] described by Captain Turner, in his journey through Bootan, ('Embassy to Tibet,' p. 160).

"Herds of this species of cattle have been long kept by many gentlemen in the eastern districts of Bengal, and also in other parts of this province; but no detailed account of the animal and of its habits has been yet published in India. To remedy this deficiency, Dr. Roxburgh undertook, at my solicitation, to describe the Gayal, from those seen by him in a herd belonging to the Governor-General. Dr. Buchanan has also obligingly communicated his observations on the same cattle; with information obtained from several gentlemen at Tipura, Sylhet, and Chatgaon, relative to the habits of the animal. The original drawing from which the plate has been taken was drawn by a native artist."

Reduced copy of the Plate just referred to.

This representation does not appear to have been taken from a specimen of the animals here described: [Pg 59] it bears a much stronger resemblance to our figure of the Gaur, which was taken from the stuffed specimen in the British Museum (see p. 97), than it does to the Gyall (*Bos frontalis* of Lambert, see p. 51), or to the Gayal, which died in the Zoological Gardens in 1846, from which our figure was taken, which is given on p. 68.

Dr. Roxburgh, who undertook, at the solicitation of Mr. Colebrooke, to describe the Gayal, appears to have done so by the very simple method of copying Mr. Macrae's description of the Gyall, which appeared in the 'Linnean Transactions,' in 1804, to which he has added, that the dewlap is deep and pendant; and this, according to every other account, is not the fact.

With respect to the account given by Dr. Buchanan, I have thought it best to quote it in full; because (although it repeats several of the characteristics already given,) it appears to flow from the pen of one who really observed what he describes.

He says: "The Gayal generally carries its head with the mouth projecting forward, like that of a Buffalo. The head, at the upper part, is very broad and flat, and is contracted suddenly towards the nose, which is naked, like that of the common cow. From the upper angle of the forehead proceed two thick, short, horizontal processes of bone, which are covered with hair; on these are placed the horns, which are smooth, shorter than the head, and lie nearly in the plane of the forehead. They diverge outward, and turn upward with a gentle curve. At the bases they are very thick, and are slightly compressed, the flat side being toward the front and the tail. The edge next the ear is rather the thinnest, so that a [Pg 60] transverse section would be somewhat ovate. Toward their tips the horns are rounded, and end in a sharp point. The eyes resemble those of the common Ox; the ears are much longer, broader, and blunter than those of that animal.

"The neck is very slender near the head, at some distance from which a dewlap commences, but this is not so deep, nor so much undulated as in the Zebu or Indian Ox. The dewlap is covered with strong longish hairs, so as to form a kind of mane on the lower part of the neck; but this is not very conspicuous, especially when the animal is young.

"In place of the hump (which is situated between the shoulders of the Zebu) the Gayal has a sharp ridge, which commences on the hinder part of the neck, slopes gradually up till it comes over the shoulder-joint, then runs horizontally almost a third part of the length of the back, where it terminates with a very sudden slope. The height of this ridge makes the neck appear much depressed, and also adds greatly to the clumsiness of the chest, which, although narrow, is very deep. The sternum is covered by a continuation of the dewlap. The rump, or os sacrum, has a more considerable declivity than that of the European Ox, but less than that of the Zebu.

"The tail is covered with short hair, except near the end, where it has a tuft like that of the common Ox; but in the Gayal the tail descends no lower than the extremity of the tibia.

"The legs, especially the fore ones, are thick and clumsy. The false hoofs are much larger than those of the Zebu. The hinder parts are weaker in proportion than the fore; and, owing to the contraction of the belly, [Pg 61] the hinder legs, although in fact the shortest, appear to be the longest.

"The whole body is covered with a thick coat of short hair, which is lengthened out into a mane on the dewlap, and into a pencil-like tuft on the end of the tail. From the summit of the head diverges, with a whirl, a bunch of rather long coarse hair, which lies flat, is usually lighter-coloured than that which is adjacent, and extends towards the horns and over the forehead. The general colour of the animal is brown, in various shades, which very often approaches to black, but sometimes is rather light. Some parts, especially about the legs and belly, are usually white; but in different individuals these are very differently disposed."

The following is the measurement of a full-grown cow:—

	Ft.	In.
From nose to summit of head	1	6
Between roots of horns	0	10
From horns to shoulder	3	3
From shoulder to insertion of tail	4	3
Height at shoulder	4	9
Height at loins	4	4
Depth of chest	2	9
Circumference of chest	6	7
Circumference at loins	5	10
Length of horns	1	2
Length of ears	0	10

"The different species of the Ox kind may be readily distinguished from the Gayal by the following marks; the European and Indian oxen by the length of their tails, which reach to the false hoofs; the American Ox, by the gibbosity on its back; the *Bovis moschatus*, Caffer, and *pumilus*, by having their horns approximated at their [Pg 62] bases; the *Bos grunniens* by it's whole tail being covered with long silky hairs; the *Bos bubalus*,(at least the Indian buffalo,) by having the whole length of its horns compressed, and by their being longer than the head, and wrinkled—also by its thin coat of hair, by its want of a dewlap, and above all by its manners; the *Bos barbatus*, by the long beard on its chin.

"The cry of the Gayal has no resemblance to the grunt of the Indian Ox, but a good deal resembles that of the Buffalo. It is a kind of lowing, but shriller, and not near so loud as that of the European Ox. To this, however, the Gayal approaches much nearer than it does to the Buffalo."

Mr. Macrae, who furnished the account in 1804, is again consulted; and from his second account, the following additional particulars have been gleaned. [Now, however, as the reader will observe, the name is Gayal, and not Gyall; although, according to Mr. Macrae's own derivation of the word, it would appear to be more correctly Gyall.]

"The Gayal is found wild in the range of mountains that form the eastern boundary of the provinces of Aracan, Chittagong (Chatgaon), Tipura, and Sylhet.

"The Cucis, or Lunclas, a race of people inhabiting the hills immediately to the eastward of Chatgaon, have herds of the Gayal in a domesticated state. By them he is called Shial, from which, most probably, his name of Gayal [Gyall] is derived; as he is never seen on the plains, except when he is brought there. It appears, however, that he is an animal very little known beyond the limits of his native mountains, except by the inhabitants of the provinces above mentioned. [Pg 63]

"His disposition is gentle: even when wild in his native hills, he is not considered to be a dangerous animal; never standing the approach of man, much less bearing his attack.

"To avoid the noon-day heat, he retires to the deepest shade of the forest; preferring the dry acclivity of the hill to repose on, rather than the low swampy ground below; and never, like the Buffalo, wallowing in mud.

"Gayals have been domesticated among the Cucis from time immemorial; and without any variation in their appearance from the wild stock. No difference whatever is observed in the colour of the wild and tame breeds; brown of different shades being the general colour of both.

"The wild Gayal is about the size of the wild Buffalo of India. The tame Gayals among the Cucis, being bred in nearly the same habits of freedom, and on the same food, without ever undergoing any labour, grow to the same size with the wild ones.

"The Cucis makes no use whatever of the milk, but rear the Gayals entirely for the sake of their flesh and skins; they make their shields of the hides of these animals. The flesh of the Gayal is in the highest estimation among the Cucis; so much so, that no solemn festival is ever celebrated without slaughtering one or more Gayals, according to the importance of the occasion.

"The domesticated Gayals are allowed by the Cucis to roam at large during the day, through the forest, in the neighbourhood of the village; but as evening approaches, they all return home of their own accord; the young Gayal being early taught this habit, by being regularly fed every night with salt, of which he is very fond; and [Pg 64] from the occasional continuance of this practice, as he grows up, the attachment of the Gayal to his native village becomes so strong, that when the Cucis migrate from it, they are obliged to set fire to the huts which they are about to leave, lest their Gayals should return thither from their new place of residence, before they become equally attached to it, as to the former, through the same means.

"The wild Gayal sometimes steals out from the forest in the night, and feeds in the rice fields bordering on the hills. The Cucis give no grain to their cattle. With us (at Chatgaon) the tame Gayals feed on Caláï *(phaseolus max)*; but as our hills abound with shrubs, it has not been remarked what particular kind of grass they prefer.

"The Hindus in this province will not kill the Gabay (or Gayal) which they hold in equal veneration with the cow. But the As'l Gayal, or Seloï, they hunt and kill, as they do the wild Buffalo. The animal here alluded to is another species of Gayal found wild in the hills of Chatgaon. He has never been domesticated, and is in appearance and disposition very different from the common Gayal which has just been described. The natives call him the As'l Gayal, in contra-distinction to the Gabay. The Cucis distinguish him by the name of Seloï; and the Mugs and Burmas by that of P'hanj, and they consider him, next to the tiger, the most dangerous and fiercest animal of their forests."

Mr. Elliot, in writing from Tipura, says,—"I have some Gayals at Munnamutty, and from their mode of feeding I presume that they keep on the skirts of the vallies, to enable them to feed on the sides of the [Pg 65] mountain, where they can browse; they will not touch grass, if they can find shrubs.

"While kept at Camerlah, which is situated in a level country, they used to resort to the banks, and eat on the sides; frequently betaking themselves to the water, to avoid the heat of the sun. However, they became sickly and emaciated, and their eyes suffered much; but, on being sent to the hills, they soon recovered, and are now (1808) in a healthy condition. They seem fond of the shade, and are observed in the hot weather to take the turn of the hills, so as to be always sheltered from the sun. They do not wallow in mud, like Buffaloes, but delight in water, and stand in it during the greatest heat of the day, with the front of their heads above the surface.

"Each Cow yields from two and a half to about four sérs [from five to eight pounds] of milk, which is rich, sweet, and almost as thick as cream; it is of a high flavour, and makes excellent butter."

We learn from Mr. Dick that the Gayal is called Gaujangali in the Persian language, Gavaya in Sanscrit, and Mat'hana by the mountaineers; but others name the animal Gobay-goru.

The tame Gayals, however long they may have been domesticated, do not at all differ from the wild ones, unless in temper, for the wild ones are fierce and untractable. The colour of both is the same, namely, that of the Antelope, but some are white and others black, none are spotted or piebald. They graze and range like other cattle, and eat rice, mustard, chiches, and any cultivated produce, as also chaff and chopped straw.

According to this gentleman the Gayal lives to the age of twenty or twenty-five years, and reaches its full growth [Pg 66] at five years. The female is generally higher than the male. She receives the bull in her fifth year, and bears after ten months.

In reference to the case of Mr. Bird's Gayal breeding with the common Zebu, I may observe that this proves nothing beyond the bare fact stated; no inference whatever of an identity of species can be drawn from a thousand such cases. It is pretty well known that animals of perfectly distinct species will, when artificially brought together, produce hybrids, as in the familiar examples of the Horse and the Ass, the Canary and the Goldfinch; but a hybrid is neither a species nor (zoologically speaking) a variety.

In a paper on the Gour, by General Hardwicke, ('Zoological Journal,' Vol. III,) he introduces the following observations on the Gayal: "Of the Gayal (*Bos Gavæas* of Colebrooke) there appears to be more than one species. The provinces of Chatgong and Sylhet produce the wild, or, as the Natives term it, the Asseel Gayal, and the domesticated one. The former is considered an untameable animal, extremely fierce, and not to be taken alive. It rarely quits the mountain tract of the south-east frontier, and never mixes with the Gobbay, or village Gayal of the plains. I succeeded in obtaining the skin, with the head, of the Asseel Gayal, which is deposited in the Museum of the Hon. East-India Company, in Leadenhall Street." [A drawing was taken of this head, of which the engraving on the opposite page is a copy.]

"I may notice another species of Gayal, of which a male and female were in the Governor General's park, at Barrackpore. This species differs in some particulars [Pg 67] from the domesticated Gayal, and also from the Asseel, or true Gayal; first, in size, being a larger animal than the domestic one; secondly, in the largeness of the dewlap, which is deeper and more undulated than in either the wild or tame species; and, thirdly, in the size and form of the horns."

Thus, according to the opinion of General Hardwicke, there are three distinct species of the Gayal; but in this matter nothing can be decided without further evidence, which we hope will soon appear in the shape of complete skeletons, and accurate drawings and descriptions.

[Pg 68]

THE TAME OR DOMESTIC GAYAL.

The representation of the Gayal here given was taken from a living specimen in the Zoological Gardens, 1846.

The scanty information I was able to glean concerning it, consists in its having been procured at Chitagong, and shipped, as a commercial speculation, from Calcutta for London, in January 1844, when about two years and a half old. It remained in the Zoological Gardens till the summer of 1846, when it died from inflammation of the bowels, brought on chiefly by eating too much green food.

I had the above particulars from Mr. Bartlett, naturalist, &c., who had been commissioned to dispose of it. He preserved the skeleton, which he kindly allowed me to examine, and from which I made the sketches of the skull and horns, which appear on the following page.

The skeleton has fourteen pairs of ribs. [Pg 69]

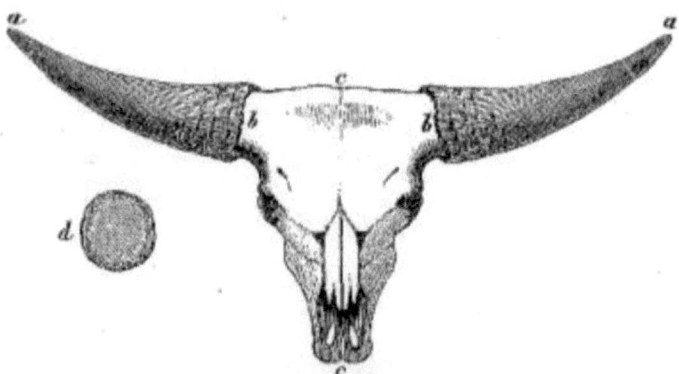

Skull of Domestic Gayal, viewed in front, with Section of Horn.

	Inches.
Distance from tip to tip (a to a)	39
Length of horn (a to b)	16
Circumference of horn at base	17
Distance of bases (b to b)	11
Length of skull (c to c)	19

Fig. d, section of the horn, at the base.

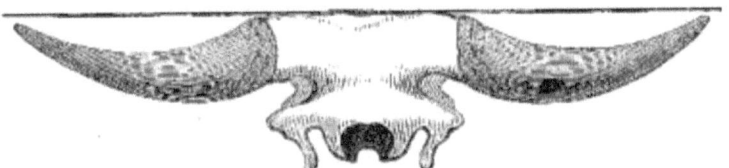

Occipital view of the same Skull.

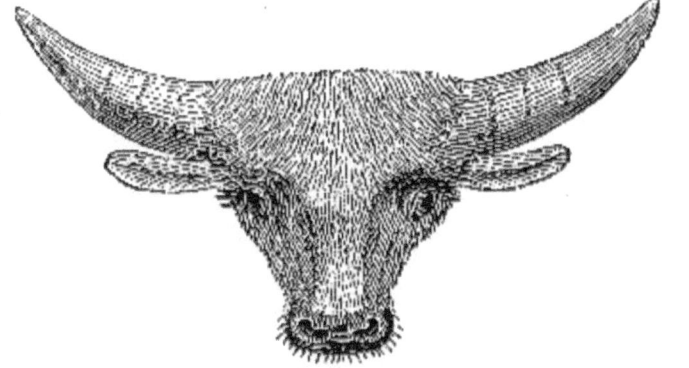

Head of Domestic Gayal.

[Pg 70]

In concluding these details of the Gayal and Gyall, let it be remarked that, when we hear one animal called Gayal and another Gyall, we are not, *on that account merely*, to set them down as of the same species. It is hardly necessary to say, that similarity or even identity of name, is not the slightest criterion of identity of species. The name Elephant is popularly applied to that animal, whether brought from Africa or Asia; they are, nevertheless, anatomically distinct. The same observation may be made respecting the Lions of those countries, and various other animals.

It may further be observed, that the value of external characters in determining a species is very different when applied to ascertain the distinctions of domestic races, to what it is when applied to ascertain the distinctions of animals living in a natural state. In domestication, varieties ramify to an indefinite extent, and under such circumstances external characters are comparatively valueless. But wild animals retain their external characters with undeviating exactness; exceptional cases may indeed occur, but so very rarely, that they are not worth taking into the account; consequently, external forms, and in some cases even colours, become of importance in ascertaining specific distinction.

[Pg 71]

THE JUNGLY GAU.

Bos Sylhetanus. **(Cuv.)**

Further information is requisite to decide the specific character of this animal. According to the opinion of Col. Smith, (see 'Synopsis of the Species of Mammalia' in Griffith's Translation of Cuvier's Animal Kingdom,) it is a mere variety of the Gayal (*Bos Gavæus*); and Mr. J. E. Gray, in his 'List of the Specimens of Mammalia in the Collection of the British Museum,' classes it as a domestic variety of the same animal, but Mr. Fred. Cuvier regards it as an entirely new species.

The following account of the Jungly Gau (which is the only one that has been published), is a translation from the splendid folio work of Messrs. St. Hilaire and F. Cuvier.

This species of Ox, which is entirely new, appears to [Pg 72] be the most nearly allied to our domestic cattle. Those ruminants which are classed under the generic name of Ox, may be very naturally divided into two distinct groups. The first includes the Buffaloes, animals in some measure aquatic, living in low, swampy localities, or near rivers, in which they remain half immersed a great part of the day; having broad-based horns, partly spreading over their foreheads, flat on their internal side, and round on their external; tongue soft, &c. The second is that of the Ox, properly so called. These are distinguished from the first by their dwelling on more elevated lands, or in the vicinity of forests; having smooth round horns, without enlargement at their base; tongue covered with horny papillæ, &c.

It is to this second family, consisting of the American Bison, the Aurox, the Yak, and the domestic Ox, with its varieties, that the Jungly Gau undoubtedly belongs. It however differs from the first two in being entirely destitute of the thick shaggy mane; and, instead of the long silky hair of the third, it is clothed with close, short hair, equal in uniformity of texture to the sleekest of our domestic cattle. To judge from its general appearance, we might be even tempted to take it for a mere variety of the domestic species, so close is the resemblance. But the information furnished by M. Alfred Duvaucel, in the only description which has been given, leaves no doubt as to its being a new species.

The following is M. Duvaucel's account:—"The horns of the Jungly Gau rise from the sides of the occiput, first outward, then forward, with a slight inclination backward of the upper extremity, forming a double lunation, and separated by a space which gradually diminishes as the [Pg 73] animal grows older; standing equally apart in every individual of the same age and sex; are round, except at their base, which is slightly compressed; and they become smoother as the animal advances in age.

"The hump, which is characteristic of the generality of Indian oxen, is reduced in this to a slight prominence, extending to the middle of the back, and is covered with a grayish, woolly hair, rather longer than that on the other parts of the body, which spreads likewise over the occiput and the front. The rest of the hair is black except the legs, which are white from the knees downwards. The tail terminates in a large tuft of hair; and, in bulls of two or three years old, the under part of the neck is slightly furnished with long, black, silky hair.

"The female is smaller than the male, with horns of a still less proportionate size. The front of the head, instead of being convex, as in the male, appears to be slightly depressed, in consequence of the superior elevation of the muzzle. The colour of the female is not so deep a black; the gray on the top of the neck and the shoulders extends to the sides, and the inferior part of the muzzle is white.

"I have long entertained the opinion," continues M. Duvaucel, "that these oxen were essentially the same as the domestic—that they were both varieties of the same species; but this opinion was formed on the inspection only of such specimens as I had seen in the menagerie at Barracpour. Since that time, I have pursued them myself near the mountains of Sylhet; and I have likewise learned from various sources that they are as numerous and as generally diffused as the common Buffalo; but they appear to be wilder than the Buffalo, and not so [Pg 74] bold, never approaching where man has established his dominion. Nevertheless, when caught, they are easily subdued, and become quite domesticated in a few months. The milk of this species is said to be more abundant and nourishing than that of any other."

From all that is at present known respecting this animal, it is regarded by M. F. Cuvier as a new species added to the genus *Bos*; and, from the circumstance of its having been first seen in a wild state near the mountains of Sylhet, he has given it the specific name of *Sylhetanus*.

The animal represented in the following vignette is the Syrian Ox, which is considered as a variety of *Bos Taurus*.

THE BUFFALO.

The animal generally known under the name of the *Common* Buffalo is evidently a different species from the *Cape* Buffalo. Much confusion, however, prevails in the accounts, both of travellers and naturalists, on the subject of these two animals. Descriptions of the one are mingled with descriptions of the other, and anecdotes are related of the one which, there is good reason for believing, ought to be referred to the other. It is highly probable that future and more accurate observations will show that more than one species has been confounded under the general epithets of "the common Buffalo," "the domestic Buffalo," "the tame Buffalo," or, more indeterminate still, "*the* Buffalo."

The accounts furnished by travellers of the various animals in Asia and Africa, described by them as Buffaloes, are altogether vague and unsatisfactory, and frequently erroneous; not from any desire on the part of the authors to deceive, but merely because their observations have been made in the most careless and indifferent manner; and, in many instances, their information is obtained from the verbal communications of ignorant natives.

In those descriptions which are confined to the Buffalo, as it at present exists in Italy and the south of Europe, tolerable reliance may be placed, as their character and habits are there well known, being of every day observation; yet, even in this case, little or nothing is known [Pg 76] of the anatomy of the animal, and its period of gestation has never been precisely stated. The following information on this latter point is given in Griffith's 'Cuvier,' (vol. iv, p. 383,) "Gestation *is said* to last twelve months, but *it appears* not to exceed ten."

THE ITALIAN BUFFALO.

Bos Bubalus.

This animal is more bulky than the domestic Ox, and its limbs are stouter. The head is larger, in proportion to the size of the body, than that of the domestic Ox, and is generally carried with the muzzle projecting; the forehead is rather convex, and higher than broad; the horns are large, slightly compressed, and recline towards the neck, with the points turned up; dewlap of a moderate size. [Pg 77]

Throughout the whole range of the Italian peninsula Buffaloes are used as beasts of burden, and their immense strength renders their services invaluable in the marshy and swampy districts, where the services of horses, or ordinary oxen, would be totally unavailing. The roads through which they are obliged to pass are frequently covered to a depth of two or three feet, through which they work their way with wonderful perseverance.

On the great plain of Apulia the Buffalo is the ordinary beast of draught; and at the annual fair held at Foggia, at the end of May, immense droves of almost wild Buffaloes are brought to the town for sale. Fearful accidents occasionally happen; enraged animals breaking from the dense mass, in spite of all the exertions of their drovers, and rushing upon some object of their vengeance, whom they strike down, and trample to death. It is dangerous to overwork or irritate the Buffalo, and instances have been known in which, when released by the brutal driver from the cart, they have instantly turned upon the man and killed him on the spot.

The following part of their history is remarkable: They appear to be most numerous, and to thrive best in those districts which are most infected with malaria. In the Pontine marshes they find a favorite retreat, and in the pestilential Maremma scarcely any other animals are to be seen. In the northern portions of Italy, where malaria is much less frequent than in the south. Buffaloes are to be found in the greatest numbers precisely in those localities where malaria is the most prevalent.

They are particularly fond of the long rank herbage, which springs up in moist and undrained lands. In their [Pg 78] habits they are almost amphibious, lying for hours half submerged in water and mud.

When travellers make use of the name "common Buffalo," they are usually understood to mean an animal identical with the Italian species; if this really be the case, its geographical range must be very extensive. It is said to inhabit the extensive regions of Hindostan, China, Cochin-China, Malabar, Coromandel, Persia, and the Crimea; also Abyssinia, Egypt, and the south of Europe; to which may be added, most of the large islands in the Indian Sea.

As an article of food, the flesh of this animal is inferior to the beef of the domestic Ox, but the milk of the female is particularly rich and abundant; the semi-fluid butter, called *ghee* in India, is made from it. According to the testimony of Colonel Sykes, the long-horned variety is reared in vast numbers in the Mawals, or hilly tracts lying along the Ghauts:—"In those tracts much rice is planted, and the male Buffalo, from his superior hardihood, is much better suited to resist the effects of the heavy rains, and the splashy cultivation of the rice than the bullock. The female is also infinitely more valuable than the cow, from the very much greater quantity of milk she yields." The hide is also much valued for its strength and durability.

In India they are used as beasts of burden; but the nature of the goods they carry must be such as will not suffer from being wet, as they have an invincible propensity to lie down in water. The native princes use them to fight with tigers in their public shows; and from their fierce and active nature, when excited, they frequently prove more than a match for their formidable assailants. With the native herdsman, however, they are generally [Pg 79] docile: these men ride on their favorites, and spend the night with them in the midst of jungles and forests, without fear of wild beasts. When driven along, the herds keep close together, so that the driver, if necessary, walks from the back of one to the other, perfectly at his ease. In the south of Europe they are managed by means of a ring passed through the cartilage of the nose, but in India it is a mere rope.

Their fierceness and courage are well exemplified in the following anecdote, related by Mr. D. Johnson in his interesting 'Sketches of Indian Field Sports:' "Two Biparies, or carriers of grain and merchandise on the backs of bullocks, were driving a loaded string of these animals from Palamow to Chittrah: when they were come within a few miles of the latter place, a tiger seized on the man in the rear, which was seen by a Guallah (herdsman), as he was watching his Buffaloes grazing. He boldly ran up to the man's assistance, and cut the tiger severely with his sword; upon which he dropped the Biparie, and seized the herdsman. The Buffaloes observing it, attacked the tiger, and rescued the herdsman; they tossed him about from one to the other, and, to the best of my recollection, killed him. Both the wounded men were brought to me; the Biparie recovered, and the herdsman died."

Speaking of the Buffalo at Malabar, Dillon says, "It is an ugly animal, almost destitute of hair, goes slowly, but carries very heavy burdens. Herds may be seen, as of common cows; and they afford milk, which serves to make butter and cheese. Their flesh is good, though less delicate, than that of the ox: the animal swims perfectly well, and traverses the broadest rivers. Besides the tame [Pg 80] ones, there are wild Buffaloes, which are extremely dangerous, tearing men to pieces, or crushing them with a single blow of the head; they are less to be dreaded in woods than elsewhere, because their horns often catch in the branches, and give time for the persons pursued to escape by flight. The skins of these animals serve for an infinity of purposes, and even cruses are made of them for holding water or liquors. The animals on the coast of Malabar are all wild, and strangers are not prevented from hunting them for their flesh."

Whether the animals alluded to, in all these cases, constitute only one species, or consist of several, the accounts which have been given of them (from their vagueness and want of precision) afford no means of deciding.

The following tail-piece is a representation of the Herefordshire Cow, *Bos Taurus*.

[Pg 81]

The Manilla Buffalo.

Bos Bubalis?

The animal which is represented in the above engraving, was living in the Zoological Gardens, Regent's Park, in 1846, at which time the sketch was taken.

In size the Manilla Buffalo is about equal to the Kyloe Ox. The horns are of a similar shape, and take nearly the same direction, as those of the Italian Buffalo. They differ, however, from the horns of the Italian Buffalo in three particulars: first, in not being above half so thick or bulky; second, in having a much larger curve; and third, in being considerably more compressed, which compression exists throughout their entire length: the colour of the upper surface of the horn is lightish, on the lower side nearly black. The head is narrow, and the muzzle fine; the ears are long and nearly naked; the eyes large and bright, with a peculiarly timid and suspicious expression. The limbs are slender, and indeed the whole frame is slight, and seems to betoken greater speed than strength.

We have a notable example of the uncertainty of framing generic characters, before the peculiar attributes [Pg 82] of each species are known, in Griffiths' work, already referred to (vol. iv, p. 382). "Buffaloes *in general*" are there said to possess *strong and solid* limbs, *large* head, *broad* muzzle, *long* and slender tail, back *rather* straight. Here we have an animal (a Buffalo by universal consent) whose limbs are *slender*, head *small*, muzzle *fine*; whose tail is *not* long, and whose back is any thing but straight. The Cape Buffalo, also, (see p. 86,) has *rather* a small head, its tail is absolutely *short*, and its back has very considerable curvature.

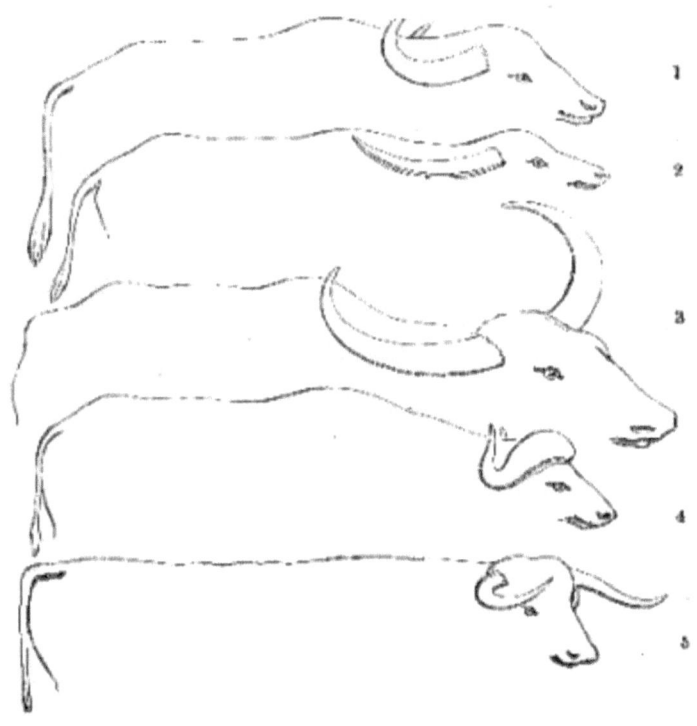

[Pg 83]

The preceding outline of the backs of four Buffaloes will show how inappropriate the character of a *straight back* is, when applied to "Buffaloes *in general*." The lowest outline (5), inserted by way of contrast, represents the back of the Domestic Ox, to which the character of straight might very properly be applied. (1) Italian Buffalo. (2) Manilla Buffalo. (3) Pulo Condore Buffalo. (4) Cape Buffalo.

Generic characters should be such (and such *only*) as will apply to every species included in the genus.

The period of gestation of the Manilla Buffalo is between forty-eight and forty-nine weeks. In two actual cases of a female now living in the Zoological Gardens, the periods were, in the one case,

340 days, in the other, 341 days; being 70 days longer than the ordinary term of the domestic Cow.

Head of Manilla Buffalo—female.

[Pg 84]

PULO CONDORE BUFFALO.

Bos Bubalus?

Not much is known of the Buffalo which is found in the island of Pulo Condore. It is related by those navigators who completed the voyage to the Pacific Ocean, begun by Captain Cook, that when at Pulo Condore, they procured eight Buffaloes, which were to be conducted to the ships by means of ropes put through their nostrils and round their horns; but when they were brought within sight of the sailors, they became so furious that some of them tore out the cartilage of their nostrils, and set themselves at liberty. All attempts to get them on board would have proved fruitless, had it not been for some [Pg 85] children, whom the animals would suffer to approach them, and by whose puerile management their rage was quickly appeased; and when the animals were brought to the beach, it was by their assistance, in twisting ropes around their legs, that the men were enabled to throw them down, and by that means get them into the boats. And what appears to have been no less singular than this circumstance was, that they had not been a day on board before they became perfectly gentle.

Whether this be a distinct species, or merely a variety, we have not, at present, the least means of ascertaining.

Osteology unknown.

Period of gestation unknown.

The tail-piece below represents a short-horned Bull of the Domestic species, *Bos Taurus*.

[Pg 86]

THE CAPE BUFFALO.

Bos Caffer.

This species of ox is only to be found in Africa, and is chiefly confined to the wooded districts lying north of the Cape of Good Hope. What Lavater endeavours to prove of the human being, namely, that the face is the index of the mind or disposition, may be applied, with at least equal truth, to the Cape Buffalo. His broad, projecting muzzle, lowering eyebrows, shaggy pendulous ears, surmounted by a pair of huge horns, give a look of bold determination to this animal, which forms a tolerably correct index of his character; his firm-set limbs and bulky body convey a no less adequate idea of his enormous strength.

These animals are gregarious, living in small herds in the brushwoods or open forests, of Caffraria, occasionally uniting in large droves. Old bulls are often met with [Pg 87] alone; but though they are fiercer than the young ones, they are less dangerous, because less active, and less inclined to exertion.

It is worthy of observation, that the males of every species of the Genus Bos are remarkably bold and courageous, as are likewise the females when they have calves. It is not, therefore, surprising that the hunting of this animal should be attended with danger, and frequently with fatal consequences. The European colonists generally pursue the sport on horseback; but the Caffers and other natives, who are more active, and accustomed to the intricacies of the forest, prefer following the game on foot.

Professor Thunberg, whilst investigating the interior of Caffraria, in 1772, in company with a sergeant and a European gardener, who had resided in the colony some time, and who acted as guide on the occasion, met with the following perilous adventure:—

"We had not advanced far into the wood," says the traveller, "before we had the misfortune of meeting with a large old male Buffalo, which was lying down quite alone, in a spot that was free from bushes for the space of a few square yards. He no sooner discovered Auge, the gardener, who went first, than, roaring horribly, he rushed upon him. The gardener turning his horse short round, behind a large tree, by that means got in some measure out of the Buffalo's sight, which now rushed straight forward towards the sergeant, who followed next, and gored his horse in the belly in such a terrible manner, that it fell on its back that instant, with its feet turned up in the air, and all its entrails hanging out, in which state it lived almost half an hour. The gardener and the sergeant, [Pg 88] in the meantime, had climbed up into trees, where they thought themselves secure. The Buffalo, after this first achievement, still appeared to take his course in the same direction, and, therefore, could not have failed in his way to pay his compliments to me, who all the while was coming towards him, and, in the narrow pass formed by the boughs and branches of the trees, and on account of the rustling noise these made against my saddle and baggage, had neither seen nor heard anything of what had passed; as in my way I frequently stopped to take up plants, and put them into my handkerchief, I generally kept behind my companions.

"The sergeant had brought two horses with him for the journey. One of them had already been despatched, and the other now stood just in the way of the Buffalo, who was going out of the wood. As soon as the Buffalo saw this second horse, he became more outrageous than before, and he attacked it with such fury, that he not only drove his horns into the horse's breast, and out again through the very saddle, but also threw it to the ground with such violence, that it died that very instant, and most of its bones were broken. Just at the moment that he was occupied with this latter horse, I came up to the opening, where the wood was so thick that I had neither room to turn my horse, nor to get on one side; I was, therefore, obliged to abandon him to his fate, and take refuge in a tolerably high tree, up which I climbed.

"The Buffalo, having finished this his second exploit, suddenly turned round, and shaped his course the same way which we had intended to take.

"From the height of my situation in the tree, I could plainly perceive one of the horses quite dead; the other [Pg 89] sprawling with his feet, and endeavouring to rise, which it had not strength to do; the other two horses shivering with fear, and unable to make their escape; but I could neither see nor hear anything of my fellow-travellers, which induced me to fear that they had fallen victims to the first transports of the Buffalo's fury. I, therefore, made all possible haste to search for them, to see if I could, in any way, assist them; but not discovering any trace of them in the whole field of battle, I began to call out after them, when I discovered these magnanimous heroes sitting fast, like two cats, on the trees, with their guns on their backs, loaded with fine shot, and unable to utter a single word.

"I encouraged them as well as I could, and advised them to come down, and get away as fast as possible from such a dangerous place, where we ran the risk of being once more attacked. The sergeant at length burst out into tears, deploring the loss of his two spirited steeds; but the gardener was so strongly affected, that he could scarcely speak for some days after."

Speaking of a small settlement in the interior, he says: "Buffaloes were shot here by a Hottentot, who had been trained to the business by the farmer, and in this manner found the whole family in meat, without having recourse to the herd. The balls were counted out to him every time he went a shooting, and he was obliged to furnish the same number of dead Buffaloes as he received of balls. Thus the many Hottentots that lived here were supported without expense, and without the decrease of the tame cattle which constitute the whole of the farmer's wealth. The greatest part of the flesh of the Buffalo falls to the share of the Hottentots, but the hide to that of the master." [Pg 90]

Young Cape Buffalo.

[Pg 91]

The Caffres, who at that time (1772) did not possess fire-arms, were, nevertheless, dextrous in the use of their javelins. When a Caffre has discovered a spot where several Buffaloes are assembled, he blows a pipe, made of the thigh-bone of a sheep, which is heard at a great distance. In consequence of this, several of his comrades run up to the spot, and surrounding the Buffaloes, at the same time approaching them by degrees, throw their javelins at them. In this case, out of ten or twelve Buffaloes, it is very rare for one to escape. It sometimes happens, however, that while the Buffaloes are running off, some one of the hunters, who stands in the way of them, is tossed and killed, which, by the people of this nation, is not much regarded. When the chase is over, each one takes his share of the game.

Since the introduction of fire-arms by the Europeans, the natives, as well as the colonists, bring down the Buffalo by means of the gun. Nevertheless, great circumspection is required in following the sport, as the animal is sometimes capable of revenging himself even after being severely wounded. On one occasion a party of huntsmen discovered a small herd of Buffaloes grazing on a piece of marshy ground. As it was impossible to get near enough without crossing a marsh, which did not afford a safe footing for their horses, they left them in charge of the Hottentots, and proceeded on foot, thinking, that if the Buffaloes should turn upon them, it would be easy to retreat by crossing the quagmire, which, though firm enough to support a man, would not bear the weight of a Buffalo. They advanced accordingly, and, under shelter of the bushes, approached with such advantage, that the first volley brought down three of the fattest of the herd, [Pg 92] and so severely wounded the great bull leader, that he dropped on his knees, bellowing most furiously. Supposing him mortally wounded, the foremost of the huntsmen issued from the covert, and began reloading his musket as he advanced, to give him a finishing shot; but no sooner did the enraged animal see his enemy in front of him than he sprang up, and ran furiously upon him. The man, throwing down his gun, fled towards the quagmire; but the beast was so close upon him, that, despairing to escape in that direction, he suddenly turned round a clump of copsewood, and began to ascend a tree. The raging animal, however, was too quick for him, and bounding forward with a tremendous roar, he caught the unfortunate man with his terrible horns, just as he had nearly escaped his reach, and tossed him into the air with such force, that the body fell dreadfully mangled into the cleft of a tree. The Buffalo ran round the tree once or twice, apparently looking for the man, until weakened with loss of blood, he again sank on his knees. The rest of the party, recovering from their confusion, then came up and despatched him, though too late to save their comrade, whose body was hanging in the tree quite dead.

The length of a full-grown Buffalo is about eight feet from horns to root of tail, and the height five feet and a half. The horns are massive and heavy, measuring from six to nine feet, following the curve from tip to tip. They are broad at the base, and very nearly meet on the centre of the forehead. Hamilton Smith says, they are "in contact at the base;" but this is not the case in the several specimens which I have examined, namely, three in the College of Surgeons, four in the British Museum, and two in the Zoological Gardens. [Pg 93]

In the living specimen in the Zoological Gardens, from which the figure at the head of this article was taken, there is a good deal of hair of a dark brown colour on the neck and shoulders, and some small tufts on the fore-legs, but the rest of the body is almost naked. The tail is short, with a tuft at the end.

The individual here referred to is by no means a large specimen, being only four feet ten inches high at the shoulders; probably he is young, and not yet full-grown. He is so active, as to be able to clear a four-feet fence, and he frequently leaps over the half-door (about three feet high,) which separates his little enclosure from his dormitory. His intelligence is much superior to that of ordinary cattle: the entrance to his apartment is furnished with four doors, two on each door-post; and when closed, they of course meet in the middle of the entrance. When he is outside, (as the doors all open inwardly,) a mere push with his horns sends them open. But when he is inside, it requires four distinct operations to shut them, and these he performs with the greatest adroitness, going from one to the other, until all are closed. He opens them also from within with equal skill, by applying the tip of one of his horns to each separately, and retiring a step or two to allow them room to open.

The flesh of the Cape Buffalo is reckoned excellent eating, especially that of the young calf, which is equal to the veal of the domestic calf. The horns are made into various articles, having a fine close grain, and taking a beautiful polish. But the hide is the most valuable part of this animal, being so thick and tough, that shields, proof against a musket-shot, are formed of it; and it affords the strongest and best thongs for harness and [Pg 94] whips. The skin of the living Buffalo is so dense that it is impenetrable, in many parts, to an ordinary musket-ball; the balls used by the huntsmen are, therefore, mixed with tin, and even these are often flattened by the resistance. In examining the skeleton of this Buffalo, the ribs are found to be remarkably strong and wide—measuring from three inches to three inches and seven-tenths in width, and overlapping each other like the scales of a fish: the difficulty of wounding this animal may be partly owing to this arrangement of the ribs.

Since the increase of the settlements about the Cape of Good Hope, the Buffalo has become rather a rare animal in the colony;

but, on the plains of Caffraria, they are so common that herds of a hundred and fifty, or two hundred, may be frequently seen grazing together towards the evening, but during the day they lie retired among the woods and thickets. They range along the eastern side of Africa, to an unknown distance in the interior.

Sparrman says that the period of gestation is twelve months.

Head of Cape Buffalo.

[Pg 95]

THE PEGASSE.

Bos Pegasus.

The above figure is copied from an engraving in the fourth volume of Griffiths' 'Cuvier,' of which the following account is given: "In the collection of drawings, formerly the property of Prince John Maurice of Nassau, now in the Berlin library, there is the figure of a ruminant with the name Pacasse written under it. Judging from the general appearance of the painting, it represents a young animal, although the horns are already about as long as the head. They are of a darkish colour, with something like ridges passing transversely, commencing at the sides of the frontal ridge, turned down and outwards, with the [Pg 96] points slightly upwards; the head is short, thick, abrupt at the nose; the forehead wide; the eyes large and full, dark, with a crimson canthus; the neck maned with a dense and rough mane; the tail descending below the hough, entirely covered with dark, long hair, appearing woolly; the carcass short, and the legs high and clumsy; but the most remarkable character appears to consist in pendulous ears, nearly as long as the head. The mane and tail are dark; the head, neck, body, and limbs dark brown, excepting the pastern joints, which are white; this figure cannot be referred to a known species, and is sufficiently curious to merit an engraving."

Swainson says that this animal only occurs in the interior of Western Africa; but he does not mention on what authority.

As the exploration of the interior of Africa is becoming an object of increasing importance and interest, we may expect, before long, to be furnished with some authentic details of the Pegasse, if such an animal really exist.

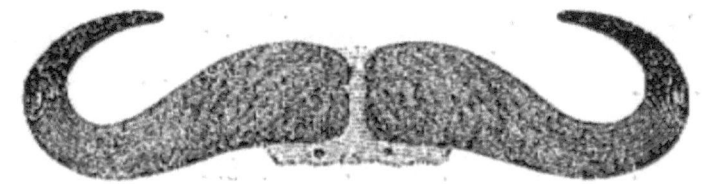

Occipital View of Horns of Bos Caffer, from a Specimen in the Zoological Society's Museum.

[Pg 97]

THE GAUR, OR GOUR.

Bos Gaurus.

The above representation of this animal was sketched from a stuffed specimen in the British Museum, the dimensions of which are given on p. 102.

The following interesting particulars are taken from Mr. T. S. Traill's paper on the Gour, in the 'Edinburgh Philosophical Journal,' October, 1824.

"The Gaur is considered by the Indians as of a species totally distinct from either the Arna or the common Buffalo. The only animal with which it appears to have affinity is the Gayal, or Bos Gavæus, described by Mr. Colebrook, in the 'Asiatic Researches,' vol. viii. That animal is said to exist, both wild and domestic, in [Pg 98] the hilly countries of Upper India, and to have a high dorsal ridge, somewhat similar to what we shall immediately find in the Gaur; but the very different form of its head, *the presence of a distinct dewlap*, and the general habit of the Gayal, appear sufficient to distinguish it from the Gaur.

The Gaur occurs in several mountainous parts of central India, but is chiefly found in Myn Pat, or Mine Paut, (Pat or Paut, in Hindostanee, signifies table-land,) a high, insulated mountain, with a tabular summit, in the province of Sergojah, in South Bahar.

This table-land is about 36 miles in length, by 24 or 25 in medial breadth, and rises above the neighbouring plains probably 2000 feet. The sides of the mountain slope with considerable steepness, and are furrowed by streams that water narrow valleys, the verdant banks of which are the favorite haunts of Gaurs. On being disturbed, they retreat into the thick jungles (of saul-trees), which cover the sides of the whole range. The south-east side of the mountain presents an extensive mural precipice from 20 to 40 feet high. The rugged slopes at its foot are covered by impenetrable green jungle, and abound with dens formed of fallen blocks of rock, the suitable retreats of Tigers, Bears, and Hyænas. The western slopes are less rugged, but the soil is parched, and the forests seem withered by excess of heat. The summit of the mountain presents a mixture of open lawns and woods. There were once twenty-five villages on Myn Pat, but they have long been deserted, on account of the number and ferocity of the beasts of prey. On this mountain, however, the Gaur maintains his seat. The Indians assert that even the Tiger has no chance in combat [Pg 99] with the full-grown Gaur, though he may occasionally succeed in carrying off an unprotected calf. The wild Buffalo abounds in the plains below the mountains; but he so much dreads the Gaur, according to the natives, that he rarely attempts to invade his haunts. The forests which shield the Gaur abound, however, in Hog-deer, Saumurs, and Porcupines.

The size of the Gaur is its most striking peculiarity. The following measurement of one not fully grown will show the enormous bulk of the animal:—

	Ft.	In.
Height from the hoof to the withers	5	11-3/4
Length from nose to end of tail	11	11-3/4

The form of the Gaur is not so lengthened as that of the Arna. Its back is strongly arched, so as to form a pretty uniform curve from the nose to the origin of the tail, when the animal stands still. This appearance is partly owing to the curved form of the nose and forehead, and still more to a remarkable ridge, of no great thickness, which rises six or seven inches above the general line of the back, from the last of the cervical to beyond the middle of the dorsal vertebræ, from which it gradually is lost in the outline of the back. This peculiarity proceeds from an unusual elongation of the spinous processes of the dorsal column. It is very conspicuous in the Gaurs of all ages, although loaded with fat; and has no resemblance to the hunch which is found on some of the domestic cattle of India. It bears some resemblance, certainly, to the ridge *described* as existing in the Gayal; but the Gaur is said to be distinguished from that animal by the remarkable peculiarity of a *total want of a dewlap*. Neither the male nor female Gaur, at any age, has the [Pg 100] slightest trace of this appendage, which is found on every other known animal of this genus.

The colour of the Gaur is a very deep brownish black, almost approaching to blueish black, except a tuft of curling dirty white hair between the horns, and rings of the same colour just above the hoof. The hair over the skin is extremely short and sleek, and has somewhat of the *oily* appearance of a fresh seal-skin.

The character of the head differs little from that of the domestic Bull, excepting that the outline of the face is more curved—the os-frontis more solid and projecting. The horns are short, thick at the base, considerably curved towards the tip, slightly compressed on one side, and in the natural state are rough. They are, however, capable of a good polish, when they are of a horn gray colour, with black solid tips. A pair in my possession measure one foot eleven inches along their convex sides; one foot from the centre of the base to the tip, in a straight line; and one foot in their widest circumference; but as they are cut and polished, a portion of their length and thickness has been lost. They are of a very dense substance, as their weight indicates, for even in their dressed state the pair weigh 5 lbs. 11 oz. avoirdupois.

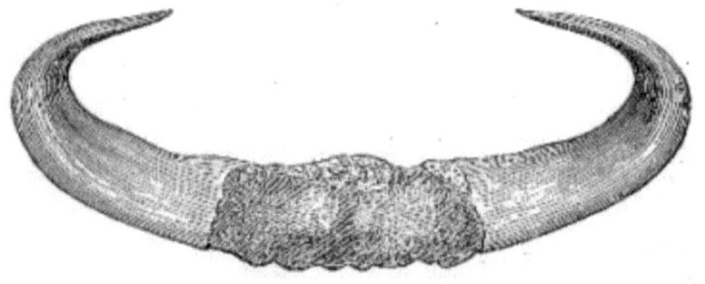

[Pg 101]

The limbs of the Gaur have more of the form of the deer than any other of the bovine genus. This is particularly observable in the acuteness of the angle formed by the tibia and tarsus, and in the slenderness of the lower part of the legs. They give the idea, however, of great strength combined with fleetness; and the animal is observed to *canter* with great velocity. The form of the hoof, too, is longer, neater, and stronger than in the ox, and the whole foot appears to have greater flexibility.

When wounded the Gaur utters a short bellow, which may be best imitated by the syllable—ugh-ugh.

It is said that the Gaur will not live in a state of captivity; even when taken very young, the calf soon droops and dies. The bull-calf of the first year is called, by the natives, Purorah; the female, Pareeah; and when full-grown the cow is called Gourin.

Gaurs associate in herds consisting usually of from ten to twenty animals. So numerous are they on Myn Pat, that, in one day hunting, the party computed that not less than eighty had passed through the station occupied by the sportsmen.

The Gaurs browse on the leaves and tender shoots of trees and shrubs, and also graze on the banks of the streams. During the cold season they remain concealed in the *saul* forests, but in hot weather come out to feed in the green vallies and lawns, which occur on the mountain of Myn Pat. They show no disposition to wallow in mire or swamps, like the Buffalo; a habit, indeed, which the sleekness of their skins renders not at all probable.

The period of gestation is said to be twelve months, and they bring forth usually in August."

To the preceding observations of Dr. Traill, I have to [Pg 102] add the important fact (which of itself will be sufficient to constitute a specific difference between the Gaur and the Gayal), namely, that in the skeleton of the Gaur there are only thirteen pairs of ribs, whilst the skeleton of the Gayal possesses fourteen pairs. This fact I have ascertained from an examination of both the skeletons; that of the Gaur in the museum of the Zoological Society, and that of the Gayal, in the possession of Mr. Bartlett, Russell Street, Covent Garden. (See p. 68.)

The skeleton of the Gaur just referred to, strikingly confirms Dr. Traill's account of the elevated dorsal ridge of this animal; several of the dorsal vertebræ measuring, with their spinous processes, upwards of seventeen inches each, the longest being twenty inches and a half.

The Gaur, from which this skeleton was taken, was killed at Nicecond, November 8, 1843. There is another fine specimen of the skull and horns of the Gaur, in the Museum of the Zoological Society, taken from an animal killed by Lieut. Nelson, on the Neilsburry Hills, Salem district. This animal measured nineteen hands and half an inch at the shoulder.

Dimensions of the Figure in the British Museum: —

	Ft.	In.
Length from nose to insertion of tail, measuring over the forehead and along the back	11	0

Height at the highest part of the dorsal ridge	5	7-1/2
Height at the croup	5	4
Length of the tail	3	1

In Mr. D. Johnson's Sketches, the Gaur is described as a kind of wild bullock, of prodigious size, residing in the Ramghur district, not well known to Europeans. Mr. Johnson says: "I have never obtained a sight of them, [Pg 103] but have often seen the print of their feet, the impression of one of them covering as large a space as a common china plate. According to the account I received from a number of persons they are much larger than the largest of our oxen; light brown colour, with short horns, and inhabit the thickest covers. They keep together in herds, and a herd of them is always near the Luggo-hill; they are also in the heavy jungles between Ramghur and Nagpoor. I saw the skin of one that had been killed by Rajah Futty Narrain; its exact size I do not recollect, but I well remember that it astonished me, having never seen the skin of any animal so large. Some gentlemen at Chittrah have tried all in their power to procure a calf without success. The Shecarries and villagers are so much afraid of these animals, that they cannot be prevailed on to go near them, or to endeavour to catch any of their young. It is a prevailing opinion in the country, that if they are in the least molested, they will attack the persons disturbing them, and never quit them until they are destroyed; and should they get into a tree, they will remain near it for many days."

The word Gau, or Ghoo, as it is sometimes spelled by European writers, appears to be used both as a generic and specific term, in Persia and Hindostan; and as it has the same meaning, and nearly the same sound, as the German word *Kuh*, and the English *Cow*, it is highly probable that its origin is the same. As the word *ur*, in Hindostan, appears to have the meaning of *wild*, or *savage*, the name Gaur, or Gau-ur, literally signifies the *wild cow*. Should the prefix *aur*, in the German word *Aurochs*, be merely a form, or different mode of spelling the prefix *ur*, then the name *Aurochs* would be precisely synonymous [Pg 104] with the Hindostanee *Gau-ur*. That *aur* is, in this instance, merely a different spelling of the prefix *ur*, would appear to be corroborated by the circumstance that the term *Urus* is the latinized form of the German *Aurochs*.—*From a MS. Note by Mr. W. A. Chatto.*

Head of Gaur, from the stuffed Specimen in the British Museum.

[Pg 105]

THE ARNEE, OR ARNA.

It does not appear, that the Arnee had been noticed by Europeans until the year 1792, when the following detailed account appeared in a weekly Miscellany, called *'The Bee,'* conducted by Dr. J. Anderson.

This animal is hitherto unknown among the naturalists of Europe. It is a native of the higher parts of Hindostan, being scarcely ever found lower down than the Plains of Plassy, above which they are found in considerable numbers, and are well known by the natives. [Pg 106]

The figure, which is given at the end of this article, is copied from a curious Indian painting, in the possession of Gilbert Innes, of Stow. It forms one of a numerous group of figures, represented at a grand Eastern festival. There are two more of them in the same painting. In this and both the others, the horns bend inwards in a circular form; and it would seem, too, that if a transverse section of the horn was made at any place, that also would be circular. But this is a defect in the painting, for although all the horns of the Arnee tribe bend in a circular form, yet if the horn be cut transversely, the section is not circular, but rather of a triangular shape. The horns of the Arnee rise in a curve upwards, nearly in the same plane with the forehead, neither bending forward nor backward. That part of the horn which fronts you when the animal looks you in the face, is nearly flat, having a ridge projecting a little forward all along, nearer the outer curvature of the horn; from that ridge outward it goes backward, not at right angles, but bending a little outward; and near the back part there is another obtuse rounded ridge, where it turns inward, so as to join another obtuse, rounded angle, at the inner curvature of the horn. Along the whole length, especially toward the base of the horn, there are irregular transverse dimples, or hollows and rugosities, more nearly resembling those of a ram, than that of a common ox's horn, but no appearance of rings, denoting the age of the animal, as in the horns of our cattle.

This description of the horns is taken from a pair of real horns of the animal, now in the possession of Mr. James Haig, merchant in Leith, that were sent home to him this year (1792) by his brother, Mr. W. Haig, of [Pg 107] the 'Hawkesbury' East-Indiaman, and of which the following cut represents a front view. The little figure marked *a*, represents a section of the horn near its base.

(1).—Horns of young Arnee—Scale of Half an Inch to a Foot.

In this young specimen (1) the length of the skull is exactly two feet, and the distance between the tops of the horns thirty-five inches. In the following sketch (2) from the Museum of the College of Surgeons, the length of the skull is likewise two feet, and the distance between the tips of the horns three feet four inches and a half.

The young animal just referred to, was found in a situation near which no other animal of this sort had ever before been discovered: it was killed by the crew of the 'Hawkesbury,' in the river Ganges, about fifty miles below Calcutta, at the place where the ships usually lie.

The flesh was eaten by the ship's company, by whom it was considered very good meat. Although conjectured to be only two years old, it weighed, when cut up, 360 lbs. the quarter, which is 1440 lbs. the carcase, exclusive of head, legs, hide, and entrails. [Pg 108]

(2).—Horns of Arnee.—Scale of Half an Inch to a Foot.

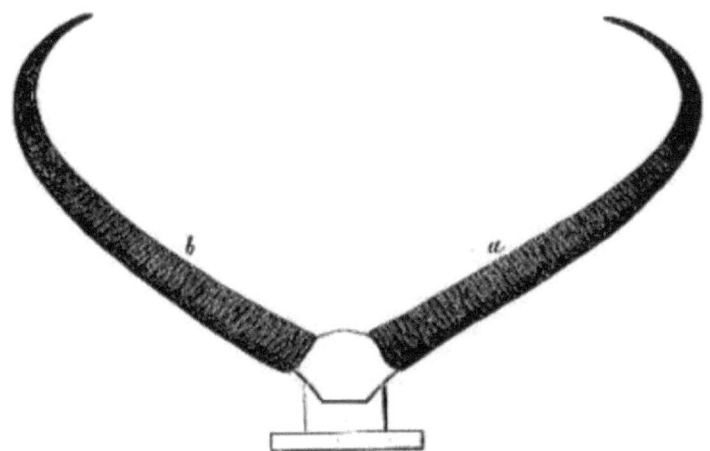

(3).—Horns of Arnee.—Scale of Half an Inch to a Foot.

[Pg 109]

This last sketch (3) is from a pair of horns in the British Museum, of which the following are the dimensions:—

	Ft.	In.
The horn *a*, from tip to base, along the outer curve	6	6
The horn *b* ditto ditto	6	3
Circumference at the base of horn *a*	1	5
Ditto ditto of horn *b*	1	6 [A]

The Arnee is by far the largest animal of the Ox tribe yet known. In its native country *it is said* to measure usually twelve, sometimes fourteen, feet from the ground to the highest part of the back! The one in the vignette, p. 111, comparing it with the man on its back, would not seem to be quite so tall.

From the appearance of the three Arnees in the painting before mentioned, it would seem that they are quite docile, and easily tamed; for they are all standing quietly, with a person on their back, who guides them by means of a rein, formed of a cord fastened to the gristle of the nose, in the Eastern manner. The colour of the animal, in all the three figures, is a pure black, except between the horns, where there is a small tuft of longish hair of a bright red colour.

From the accounts of more recent travellers, there seem to be two or three varieties of this animal, which exist, both in a wild and domestic state, in China as well as India.

According to Major Smith, the gigantic or Taur-elephant Arnee, appears to be rare; found only single, or in small families, in the upper eastern provinces and [Pg 110] forests at the foot of the Himalaya. A party of officers of the British Cavalry, stationed in the north of Bengal, went on a three months' hunting expedition to the eastward, and destroyed in that time forty-two Tigers, and numerous wild Buffaloes, but only one Arnee. When the head of this specimen rested perpendicularly on the ground, it required the out-stretched arms of a man to hold the points of the horns. These are described as angular, with the broadest side to the rear; the two others anterior and inferior; they are of a brownish colour, and wrinkled; standing outwards, and not bent back; straight for near two thirds of their length, then curving inwards, with the tips rather back. The face is nearly straight, and the breadth of the forehead is carried down with little diminution to the foremost grinder.

There is a spirited figure of a long-horned Buffalo in Captain Williamson's 'Oriental Field Sports,' which Major Smith considers to be a representation of the great Arnee; and of which Captain Williamson relates the following anecdote:—

"The late Dr. Baillie, who was a very keen and capable sportsman, used, in my idea, to run many very foolish risks among Buffaloes. I often remonstrated with him on his temerity, but he was so infatuated, that it was all to no purpose. One morning, as we were riding on the same elephant to the hunting-ground, to save our horses as much as possible, we saw a very large Buffalo lying on the grass, which was rather short and thin; as usual, the doctor would have a touch at him, and, heedless of my expostulation, dismounted with his gun. The Buffalo, seeing him approach, rose and shook his head as a prelude to immediate hostilities. My friend fired, and hit him on the side. [Pg 111] The enraged brute came thundering at the doctor, who lost no time in running round to the opposite side of the elephant; the *mohout*, at the same time, pushed forward, to meet and screen him from the Buffalo, which absolutely put his horns under the elephant's belly, and endeavoured to raise him from the ground. We had no other gun, and might, perhaps, have felt some more severe effects from the doctor's frolic, had not the Buffalo, from loss of blood, dropped at our side. The Buffalo was upwards of six feet high at the shoulder, and measured nearly a yard in breadth at the chest. His horns were above five feet and a half in length."

In systems of classification, even of very recent date, the Arnee is considered merely as a variety of the Buffalo. It appears to me, however, that our information on the subject is not yet sufficiently precise to determine this point.

FOOTNOTES:

[A] In Shaw's 'Zoology,' it is mentioned that a Mr. Dillon saw some horns in India which were ten feet long.

[Pg 112]

THE ZAMOUSE, OR BUSH COW.

Bos Brachyceros.

[The following extract, from the 'Annals of Nat. Hist.,' vol. ii, p. 284, is from the pen of Mr. J. E. Gray.]

"Captain Clapperton and Colonel Denham, when they returned from their expedition in Northern and Central Africa, brought with them two heads of a species of Ox, covered with their skins. These heads are the specimens which are mentioned in Messrs. Children and Vigors' accounts of the animals collected in the expedition, as [Pg 113] belonging to the Buffalo, *Bos Bubalus*, and they are stated to be called *Zamouse* by the natives; but, as no particular locality is given for the head, this name is probably the one applied to the common Buffalo, which is found in most parts of North Africa.

"Having some years ago compared these heads with the skull of the common Buffalo, *Bos Bubalus*, and satisfied myself, from the difference in the form and position of the horns, that they were a distinct species, in the 'Mag. of Nat. Hist.,' for 1837 (new series, vol.

i, p. 589), I indicated them as a new species, under the name of *Bos Brachyceros*.

"In the course of this summer (1838), Mr. Cross, of the Surrey Zoological Gardens, received from Sierra Leone, under the name of the *Bush Cow*, a specimen which serves more fully to establish the species. It differs from the Buffalo and all other oxen in several important characters, especially in the large size and particular bearding of the ears, and in being totally deficient in any dewlap. It also differs from the Buffalo in its forehead, being flatter and quite destitute of the convex form which is so striking in all the varieties of that animal.

"Mr. Cross's cow is, like the head in the Museum, of a nearly uniform pale chesnut colour. The hair is rather scattered, and nearly perpendicular to the surface of the body. The legs, about the knees and hocks, are rather darker. The ears are very large, with two rows of very long hairs on the inner side, and a tuft of long hairs at the tips. The body is short and barrel-shaped, and the tail reaches to the hocks, rather thin and tapering, with a tuft of long hairs at the tip. The chest is rounded and rather dependent, but without the least appearance of a dewlap; and the horns nearly resemble those of the [Pg 114] Museum specimen, but are less developed, from the sex and evidently greater youth of the animal. The Rev. Mr. Morgan informs me that the animal is not rare in the bush near Sierra Leone.

"I have added a slight sketch of Mr. Cross's animal, which I hope will enable any person to distinguish this very distinct and interesting addition to the species of this useful genus."

The engraving at the head of this article is a reduced copy of Mr. Gray's figure just alluded to. The following representation of the head is from a specimen in the British Museum.

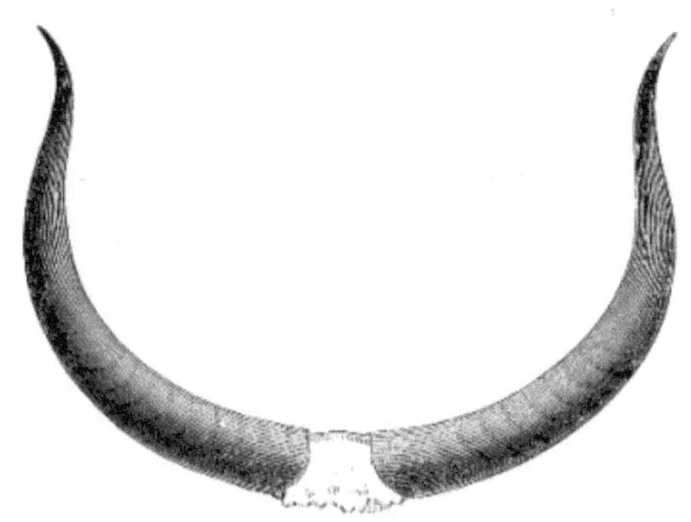

[Pg 115]

THE MUSK OX.

Bos Moschatus.

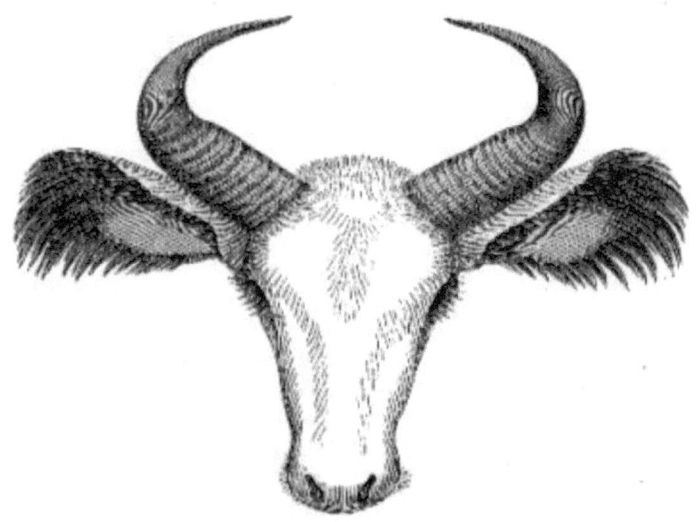

The Musk-ox, in its external appearance, more nearly resembles the Yak of Thibet than any other member of the Bos genus; and they both inhabit mountainous districts near regions of perpetual snow.

The horns of the Musk Bull are remarkably broad at their bases, which are closely united; they bend down on each side of the head, with an outward curve turning upwards towards their ends, which taper to a sharp point. They are two feet long measured along the curvature, and two feet in girth at the base; the weight of a pair of these [Pg 116] horns is sometimes sixty pounds. The broad base of the horn is hollow on the inside, and of a form approaching to a square; when this is separated from the head and the other part of the horn, it forms a convenient dish, which is very generally used by the native Esquimaux for many domestic purposes.

The horns of the cow are nine inches distant from each other at the base, and are placed exactly on the sides of the head; they are thirteen inches long, and eight or nine inches round at the base.

The head and the body generally is covered with very long silky hairs of a dark colour; some of which are seventeen inches long; on the middle of the back (which is broad and flat), the hair is lighter and not so long. Beneath the long hairs, in all parts, there is a thick coat of cinereous wool of exquisite fineness. M. Jeramie brought some to France, of which stockings were made more beautiful than silk.

The tail is only three inches long, and completely covered with very long hairs, so as to be undistinguishable to the sight. Of this tail, the Esquimaux of the northwest side of Hudson's Bay, make a cap of a most horrible appearance, for the hairs fall all round their heads, and cover their faces; yet it is of singular service in keeping off the musquitoes, which would otherwise be intolerable.

The ears are only three inches long, quite erect, and sharp pointed, but dilate much in the middle; they are thickly lined with hair of a dusky colour, marked with a stripe of white.

The frog in the hoof is soft, partially covered with hair, and transversely ribbed. The following sketch represents [Pg 117] the under surface of the foot of the Musk-ox, the external hoof being rounded, the internal pointed.

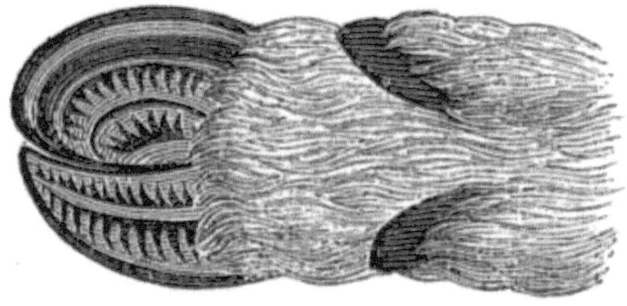

The foot-marks of the Musk-ox and those of the rein-deer are so much alike, that it requires the eye of an experienced hunter to distinguish them. The mark of the Musk-ox's hoof, however, is a little narrower.

The food of the Musk-ox is the same with that of the rein-deer — lichens and grass; and sometimes twigs and shoots of willow, birch, and pine.

At present this animal is not found in a lower latitude than 66°; but formerly they came much further to the south; and their flesh used to be brought by the natives to Fort Churchill in latitude 58°. It would appear that they are retiring northward, probably owing to the alarm created by the attacks made upon them by fire-arms. It is worthy of remark, that the American Bison has also retreated considerably to the north. According to Dr. Richardson, the Musk-ox inhabits the North Georgian Islands in the summer months. They arrive in Melville Island in the middle of May, crossing the ice from the southward, and quit it on their return towards the end of September.

The Musk-oxen, like the Bison, herd together in bands, and generally frequent barren grounds during the summer [Pg 118] months, keeping near the rivers; but retire to the woods in winter. They seem to be less watchful than most other wild animals; and when feeding are not difficult of approach, provided the hunters go against the wind. When two or three men get so near a herd as to fire at them from different points, these animals, instead of separating or running away, huddle closer together, and in this case they are easily shot down; but if the wound is not mortal, they become enraged, and dart in the most furious manner at the hunters, who must be very dexterous to evade them; for, notwithstanding the shortness of their legs, they can run with great rapidity, and climb hills and rocks, with great ease. They can defend themselves by their powerful horns against wolves and bears, which, as the Indians say, they not unfrequently kill. — (Capt. Franklin's 'Journey to the Polar Sea.')

They are hunted in their winter retreats by the Esquimaux only, the Indian tribes never visiting the barren grounds at that season.

When the Musk-ox is fat, its flesh is well tasted, and it is then preferred by the Copper Indians to the rein-deer. The flesh of bulls is high-flavoured; but both bulls and cows smell strongly of musk, their flesh at the same time being very dark and tough. The contents of the paunch, and other intestinal parts, are relished as much by the Indian as the similar parts of the rein-deer. —(Appendix to Capt. Parry's 'Second Voyage.')

The weight of the bulls killed during Capt. Parry's Second Voyage was, on an average about 700 lbs., yielding about 400 lbs. of meat. Their height, at the withers, was about ten hands and a half.

They were observed by Capt. Franklin's party to rut [Pg 119] in the end of August and beginning of September; and Hearne says, that they bring forth one calf in the latter end of May, or beginning of June; thus the period of gestation is about nine months.

The figure at the beginning of this article, as well as the following cut of the head, are from the beautiful specimen of the Musk Ox, in the British Museum.

Head of Musk Ox.

[Pg 120]

THE SANGA, OR GALLA OX.

(See Frontispiece). Bos — — ?

This singular animal is only found in Abyssinia, and is famous on account of its horns, which are of an almost incredible size.

Bruce the traveller, in speaking of these horns, says, "The animal furnishing these monstrous horns is a cow or bull which would be considered of a middling size in England. This extraordinary size of its horns proceeds from a disease that the cattle have in these countries, of which they die, and is probably derived from their pasture and climate. When the animal shows symptoms of this disorder, he is set apart in the very best and quietest grazing place, and never driven or molested from that moment. His value lies then in his horns, for his body becomes emaciated and lank, in proportion as the horns grow large; at the last period of his life, the weight of his head is so great that he is unable to lift it up, or at least for any space of time. The joints of his neck become callous at last, so that it is not any longer in his power to lift his head. In this situation he dies, with scarcely flesh to cover his bones, and it is then his horns are of the greatest value. I have seen horns that would contain as much as a common sized water-pail, such as they make use of in the houses in England." [B]

[Pg 121]

So far Mr. Bruce. Mr. Salt, who visited Abyssinia some years afterwards, gives a somewhat different account. He says: "Here [*i. e.* at Gibba], for the first time, I was gratified by the sight of the Galla Oxen, or Sanga, celebrated throughout Abyssinia for the remarkable size of its horns. Three of these animals were grazing among the other cattle in perfect health, which circumstance, together with the testimony of the natives, 'that the size of the horns is in no instance occasioned by disease,' completely refutes the fanciful theory given by Mr. Bruce respecting this creature. It appears by the papers annexed to the last edition of Mr. Bruce's work, that he never met with the Sanga; but that he made many attempts to procure specimens of the horns, through Yanni, a Greek, residing at Adowa. This old man very correctly speaks of them, in his letters, as being only brought by the Cafilas from Antalo; and I have now ascertained that they are sent to this country as valuable presents, by the chiefs of the Galla, whose tribes are spread to the southward of Enderta. So far, then, as to the description of the horns, and the purposes to which they are applied by the Abyssinians, Mr. Bruce's statements may be considered as correct; but with respect to 'the disease which occasions their size, probably derived from their pasture and climate,' 'the care taken of them to encourage this disease,' 'the emaciation of the animal,' and 'the extending of the disorder to the spine of the neck, which at last becomes callous, so that it is not any longer in the power of the animal to lift its head,' they all prove to be mere ingenious conjectures, thrown out by the author solely for the exercise of his own ingenuity.

"I should not venture to speak so positively upon this [Pg 122] matter, had I not indisputably ascertained the facts; for the Ras having subsequently made me a present of three of these animals alive, I found them not only in excellent health, but so exceedingly wild, that I was obliged to have them shot. The horns of one of these are now deposited in the Museum of the Surgeons' College, and a still larger pair are placed in the collection of Lord Valentia, at Arley Hall. The length of the largest horn of this description was nearly four feet, and its circumference at the base twenty-one inches.

"It might have been expected that the animal, carrying horns of so extraordinary a magnitude, would have proved larger than others belonging to the same genus; but in every instance which came under my observation, this was by no means the case. The etching on the following page, which was copied from an original sketch (taken from the life), may serve to convince the reader of this fact; and it will convey a better idea of the animal than any description in writing I can pretend to give. I shall only further observe, that its colour appeared to vary as much as in the other species of its genus, and that the peculiarity of the size of the horns was not confined to the male, the female being very amply provided with this ornamental appendage to her forehead."

Notwithstanding the bold and confident tone of Mr. Salt's counter-statement, it must be confessed, that the figure which he himself gives from the life (and of which the frontispiece to this volume is an exact copy), seems rather to coincide with Mr. Bruce's account, being, to all appearance, both "lank and emaciated." [Pg 123]

Engraving of the horns presented by Mr. Salt to the Museum of the College of Surgeons.

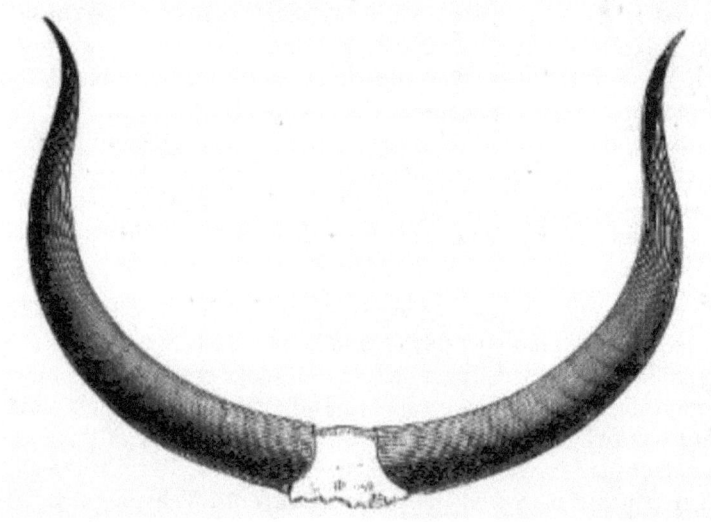

Horns of Galla Ox.

	Ft.	In.
Length of each round the outer curve	3	10-1/2
Distance between the tips	3	4
Circumference at the base	1	3
Distance between the bases at the forehead	0	3-1/2

    The Sanga is usually considered as a mere variety of *Bos Taurus*. This may possibly be the fact; but we have no proof whatever that it is so: no information on this point has been presented beyond mere conjecture. This being the case, and in the absence of direct anatomical evidence, we may be pardoned in considering it, at least, as doubtful; especially as there are so many points of external dissimilarity. The principal differences are: 1st, in the shoulder, upon which there is a hump; 2d, in the [Pg 124] back, which descends (as in the Buffaloes and Zebus), abruptly towards the tail; 3d, in the greater length of the legs; and 4th, in the forehead, which is only three inches and a half between the bases of the horns, whilst in the Common Ox it is nine inches.

The horns represented in the following sketch, are those of the Hungarian Ox (a variety of *Bos Taurus*), and are almost as remarkable for their length and expansion as those of the Abyssinian Sanga. The length of each horn is three feet four inches and a half, and the distance between the tips is five feet one inch. The sketch is from a specimen in the British Museum.

FOOTNOTES:

[B] Jerom Lobo, in his account of Abyssinia, mentions that some of the horns of the Buffaloes of that country will hold ten quarts.

[Pg 125]

INDIAN DOMESTIC CATTLE.

Bos — —?

THE ZEBU, OR BRAHMIN OX.—(*Var. a.*)

The opinions expressed in the following extract from Mr. Bennett's description of the Indian Ox (Gardens and Menag. of the Zool. Soc.), may be taken as a correct exposition of the views of naturalists generally on the subject:—

"There can be little doubt that the Zebu, or Indian Ox, is merely a variety of the Common Ox, although it is difficult to ascertain the causes by which the distinctive characters of the two races have been in the process of [Pg 126] time gradually produced. But whatever the causes may have been, their effects rapidly disappear by the intermixture of the breeds, and are entirely lost at the end of a few generations. This intermixture and its results would alone furnish a sufficient proof of identity of origin; which, consequently, scarcely requires the confirmation to be derived from the perfect agreement of their internal structure, and of all the more essential particulars of their external confirmation. These, however, are not wanting; not only is their anatomical structure the same, but the form of their heads, which affords the only certain means of distinguishing the actual species of this genus from each other, presents no difference whatever. In both the forehead is flat, or more properly slightly depressed; nearly square in its outline, its height being equal to its breadth; and bounded above by a prominent line, forming an angular protuberance, passing directly across the skull between the bases of the horns. The only circumstances in fact in which the two animals differ, consists in the fatty hump on the shoulders of the Zebu, and in the somewhat more slender and delicate make of its legs."

In a scientific work, it is not sufficient for the author merely to make an assertion; it is not even sufficient for him to say that he has made an experiment or observation, and merely give the result; he should, in every case where it is practicable, describe the nature of his experiment, — the *when*, the *where*, the *how*; — and the means and opportunity he had of making his observations, that the curious or sceptical inquirer may be enabled to perform the experiment, or make the observation for himself.

Mr. Bennett tells us, that the differences observable in the Indian Ox and the Common Ox "*rapidly* disappear [Pg 127] by the intermixture of the breeds, and are entirely lost at the end of a few generations;" but he does not refer to a single instance of this, authentic or otherwise; nor are we aware that any such instance ever occurred.

Again, he states that "their anatomical structure is the same;" but he does not inform us when, or where, or how, the comparison was made which enabled him to arrive at that conclusion.

Wishing to satisfy myself, as far as possible, on this point, I have examined the skeleton both of the British Domestic Ox and the Zebu; and the following is the result of that examination: —

NUMBER OF VERTEBRÆ.

	Cerv.	Dors.	Lumb.	Sac.	Caud.		Total.
In the Zebu	7	13	6	4	18	=	48
In the Common Ox	7	13	6	5	21	=	52

The skeletons may still be seen in the Museum of the College of Surgeons.

Furthermore, the period of gestation of the Brahmin Cow (according to the MS. records of the Zoological Society), is 300 days, while that of the Common Cow is only 270 days.

Whether the differences here pointed out are sufficient to constitute specific distinction, is left for the umpires to decide. [Pg 128]

THE ZEBU. — (Var. β.)

These Indian Cattle are extremely gentle, and admirably adapted to harness. Some of the eastern princes attach them to their artillery; but generally they employ the finest to draw their light carriages, which in form are very similar to those of the ancients. In mountainous countries, they have them shod. Their pace is a kind of amble, and they are able to sustain a journey of about twenty leagues a day. Guided by a cord which passes through the nasal cartilage, they obey the hand with as much precision as a horse.

In the same provinces are seen a race of dwarf Bisons, which are scarcely as tall as our calves of two months old, generally described under the name of *Zebu*. They are lively, well proportioned, and trained to be mounted by children, or to draw a light car. In both cases their [Pg 129] pace is a sort of amble, the same as that of the larger species.

Zebus (Var. γ) and Car.

The curious Hindoo customs in relation to this animal have been recorded by almost every traveller.

Neither the horse, the sheep, nor the goat, have any peculiar sanctity annexed to them by the Braminical superstition; it is otherwise with the cow, which in India is everywhere regarded with veneration, and is an object of peculiar worship. Representations of objects are made upon the walls with cow-dung, and these enter deeply into their routine of daily observances. The same materials are also dried, and used as fuel for dressing their victuals; for this purpose the women collect it, and bake it into cakes, which are placed in a position where they soon become dry and fit for use. The sacred character of the cow probably gives this fuel a preference to every other in the imagination of a Hindoo, for it is used in Calcutta, where wood is in abundance.

On certain occasions it is customary for the Hindoos to consecrate a bull as an offering to their deities; particular [Pg 130] ceremonies are then performed, and a mark is impressed upon the animal, expressive of his future condition to all the inhabitants. No consideration will induce the pious Bengalee to hurt or even control one of these consecrated animals. You may see them every day roaming at large through the streets of Calcutta, and tasting rice, grain, or flour in the Bazar, according to their pleasure. The utmost a native will do, when he observes the animal doing too much honour to his goods, is to urge him, by the gentlest hints, to taste of the vegetables or grain of his neighbour's stall. (*Tennant's 'Indian Recreations.'*)

One of the doctrines of the Brahmins is to believe that kine have in them somewhat of sacred and divine; that happy is the man who can be sprinkled over with the ashes of a cow, burnt by the hand of a Brahmin; but thrice happy is he who, in dying, lays hold of a cow's tail and expires with it between his hands; for thus assisted, the soul departs out of the body purified, and sometimes returns into the body of a cow. That such a favour, notwithstanding, is not conferred but on heroic souls, who contemn life, and die generously, either by casting themselves headlong from a precipice, or leaping into a kindled pile, or throwing themselves under the holy chariot wheels, to be crushed to death by the Pagods, when they are carried in triumph about the town. — (*Life of St. Francis Xavier, translated by Dryden, 1688.*)

AFRICAN AND OTHER VARIETIES.

In Shaw's Zoology, the following species or varieties are noticed:—

LOOSE-HORNED OX.

This is said to be found in Abyssinia and in Madagascar, [Pg 131] and is distinguished by pendulous ears, and horns *attached only to the skin, so as to hang down on each side*!

THE BOURY.

Of the size of a camel, and of a snowy whiteness, with a protuberance on the back, is a native of Madagascar and some other islands.

THE TINIAN OX.

Of a white colour, with black ears. Inhabits the island of Tinian.

Bewick mentions that in Persia there are many oxen entirely white, with small blunt horns and humps on their backs. They are very strong, and carry heavy burdens. When about to be loaded, they drop down on their knees like the Camel, and rise again when their burdens are properly fastened.

THE BORNOU OX,

which Col. Smith considers a distinct species, is likewise white, of a very large size, with hunched back, and very large horns, which are couched outwards and downwards, like those of the African Buffalo, with the tip forming a small half-spiral revolution. The corneous external coat is very soft, distinctly fibrous, and at the base not much thicker than a human nail; the osseous core full of vascular grooves, and inside very cellular, the pair scarcely weighing four pounds. The skin passes insensibly to the horny state, so that there is no exact demarcation where the one commences or the other ends. The dimension of a horn are:—length measured on the [Pg 132] curve, three feet seven inches; circumference at base, two feet; circumference midway, one foot six inches; circumference two thirds up the horn, one foot; length in a straight line, from base to tip, one foot five inches and a half. The species has a small neck, and is the common domestic breed of Bornou, where the Buffalo is said to have small horns.

Leguat, in his 'Voyages in 1720,' states that the oxen are of three sorts at the Cape of Good Hope, all of a large size, and very active; some have a hump on the back, others have the horns long and pendent, while others have them turned up and well shaped, as in English cattle.

Zebu. — (Var. δ.)

THE DOMESTIC OXEN OF THE HOTTENTOTS, CALLED BACKELEYS, BACKELEYERS, OR BAKELY-OSSE.

Bos — — ?

The following particulars relating to these Oxen are taken from the highly interesting work 'The Present State of the Cape of Good Hope,' by Peter Kolben, who visited that colony in 1705, and remained there during a period of eight years.

"The Hottentots have a sort of oxen they call Backeleyers, or fighting oxen; they use them in their wars, as some nations do elephants; of the taming and farming of which last creatures upon the like discipline the Hottentots as yet know nothing. They are of great use to them, too, in the government of their herds at pasture; for, upon a signal from their commanders, they will fetch in stragglers, and bring the herds within compass. They will likewise run very furiously at strangers, and therefore are of good defence against the Buschies, or robbers who steal cattle. They are the stateliest oxen of the herd: every Kraal has half-a-dozen of these oxen at the least. When one of them dies, or grows so old, that, being unfit for business, his owner kills him, a young one is chosen out of the herd to succeed him, by an ancient Hottentot, who is judged best able to discern his capacity for instruction. This young ox is associated with an old Backeleyer, and taught, by blows and other means, to follow him. At night they tie them together by the horns; and for some part of the day they fasten them together in the same manner, till at length, by this and I know not what other means, the [Pg 134] young ox is fully instructed, and becomes a watchful guardian of the herds, and an able auxiliary in war.

"The Backeleyers (so called from the Hottentot word Backeley for war) know every inhabitant of the Kraal they belong to, men, women, and children, and pay them all just the same respect that is paid by a dog to every person who dwells in his master's house. Any of the inhabitants may, therefore, at any time present themselves very safely on any side of the herds; the Backeleyers will in nowise offend them. But if a stranger, especially a European, shall approach the herds, without the company of a Hottentot of the Kraal they belong to, he must look sharp to himself; for these Backeleyers, which generally feed at the skirts of the herds, quickly discover him, and make at him upon a full gallop. And if he is not within hearing of any of the Hottentots who keep the herds, or has not a fire-arm, or a light pair of heels, or there is not a tree at hand which he can immediately climb, he is certainly demolished. The Backeleyers mind not sticks or the throwing of stones at them. This is one great reason why the Europeans always travel the Hottentot countries with fire-arms. But the first thing a European does, upon the appearance of such an enemy, is to shout and call to the Hottentots that look to the herds. The Hottentot that hears him hastens to his assistance, making all the way a very shrill whistling through his fingers. The Backeleyers no sooner hear the whistling of their keepers, which they very well know, than they stop, turn about, and return leisurely to the herds.

"But if a European, in such a case, does not (upon his shouting and calling to the keepers), hear the whistle, before the Backeleyers come up with him, he discharges [Pg 135] his fire-arm,—frightened with the report of which, the Backeleyers run away.

"I have been often run at by the Backeleyers myself. As soon as I saw them sallying out upon me, I shouted and called to the keepers. But I could not often make them hear before the Backeleyers came up with me, when I have been obliged to discharge my fire-arm (for I always carried one about with me), upon which they always turned about and left me.

"In the wars of the Hottentots with one another, these Backeleyers make very terrible impressions. They gore, and kick, and trample to death, with incredible fury. Each army has a drove of them, which they take their opportunity to turn upon the enemy. And if an army, against which the Backeleyers are sent, is not alert and upon all its guard, these creatures quickly force their way through it, tearing, shattering, and confounding all the troops that oppose them, and paving for their masters an easy way to victory. The courage of these creatures is amazing; and the discipline upon which they are formed does not a little honour to the Hottentot genius and dexterity.

"The Hottentots have likewise great numbers of oxen for carriage. These, too, are very strong and stately creatures, chosen out of the herds, at about the age of two years, by old men, well skilled in cattle. When they have destined an ox to carry burdens, they take and throw him on his back on the ground; and fastening his head and feet with strong ropes to stakes firmly fixed in the ground, they make a hole with a sharp knife through his upper lip, between his nostrils. Into this hole they put a stick, about half an inch thick, and a foot and a half long, with a hook at top to prevent its falling [Pg 136] through. By this hooked stick they break him to obedience and good behaviour; for if he refuses to be governed, or to carry the burdens they lay upon him, they fix his nose by this hooked stick to the ground, and there hold it till he comes to a better temper.

"It is an exquisite torture to an Ox to be fastened to the ground by the nose in this manner. He is not, therefore, long exercised this way, before he gets a notion of his duty, and becomes tractable. After which, the very sight alone of the stick, when he is wanton or refractory, will humble and reduce him to the will of his driver. The terror of this stick, likewise makes the carriage oxen so attentive to the words of command the Hottentots use to them, that they quickly conceive and, ever while they live, afterwards retain the intention of them. I have a thousand times been surprised at the ready obedience the carriage oxen have paid to a Hottentot's bare words. They are as quick at apprehending, and as exact in performing the orders of their driver, as is any taught dog in Europe at conceiving and accomplishing the orders of his master. The stick—the terrible stick—makes them all attention and diligence."

[Pg 137]

AFRICAN BULL.

The following notice, which will explain itself, appeared in Loudon's 'Magazine of Natural History,' for July, 1828.

"Some Account of a particular Variety of Bull (*Bos Taurus*), now exhibiting in London. By Mrs. Harvey.

"Sir,—Agreeably to your request, Mr. Harvey has taken a portrait of this animal; and as he has made the drawing on the wood himself, the engraving will be a very perfect resemblance. [C] I have, on my part, drawn up the following particulars, from what my husband told me, and I shall be happy if they prove of any interest to you or your readers:—

"This animal belongs to a French woman, who says he was brought from Africa to Bordeaux when a calf; and, after having been shown in different parts of the Continent, was taken to London, and exhibited at the Grand Bazaar in King's Street, Portman Square, last autumn. He is [Pg 138] at present five years old, four feet high at the shoulder and seven feet in length, from the horns to the insertion of the tail. The length of his face is one foot eight inches, and the girth round the collar seven feet six inches. His hair is short and silky, and the colour a cream or yellowish white, except two black tufts which appear on each foot. On the back of the neck there is a hump or swelling, which seems confined to this variety. The general aspect of the animal is mild and docile; but, when irritated, his expression is very remarkable, exhibiting itself principally in the eye. This, in its ordinary state, is very peculiar, (fig. 1, *a*,) rising more than one-half above the orbit, and bearing a resemblance to a cup and ball, thus enabling the animal to see on all sides with equal ease. The iris is naturally of a pale blue colour; but, when the animal is irritated, it varies from a very pale blue or lilac to a deep crimson. Its form is also very remarkable, being a small oval, or rather a parallelogram, with the ends cut off, and lying transversely across the ball, (fig. 1, *b*.)

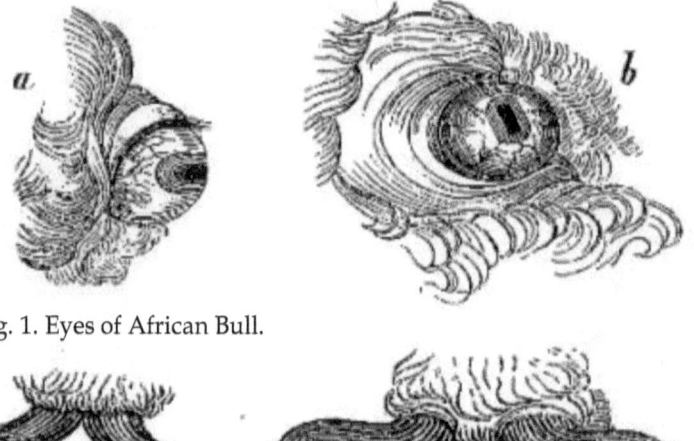

Fig. 1. Eyes of African Bull.

Fig. 2.

"The black tufts, mentioned above, are the lateral hoofs (fig. 2), which the animal sheds annually, and which grow [Pg 139] to the length of five or six inches. They are not shed together, or at stated periods; for those of the fore-feet, (*a, b,*) in this example, are at present of different ages, and, consequently, of different lengths; the difference between them being exactly that represented in the sketch.

"On the hump or collar, the hair grows much longer than on the other parts of the body, forming a sort of curled mane, resembling, I should imagine, that of the Bison. It is perfectly white, growing to the length of one foot six inches, and adding greatly to the height of the rising part behind the horns. At present the hair is only beginning to grow; but it will be in full beauty at the approach of the winter months, and will fall off gradually again in the early part of the succeeding spring.

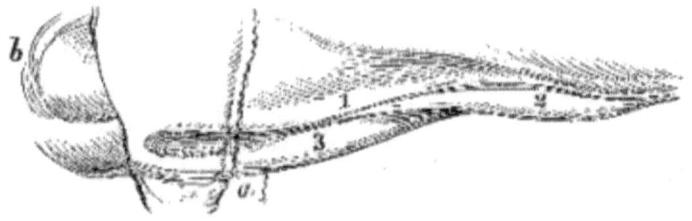

Fig. 3. Dewlap of African Bull.

"The keeper pointed out to Mr. Harvey, as a remarkable peculiarity, that the dewlap (fig. 3), in passing between the fore-legs (*a*), and under the body (*b*), seemed to divide itself into three parts, which they called the three stomachs, (1, 2, 3,) from their being very much acted on in the progress of digestion."

> I remain. Sir, &c.
> M. Harvey.

FOOTNOTES:

[C] The engraving here given as well as those of the eyes, hoofs, and dewlap, have been carefully copied from Mr. Harvey's originals.

[Pg 140]

CHILLINGHAM WHITE CATTLE.

Bos Taurus. — Restricted Variety.

Considerable interest has always been connected with the history of those herds of white cattle which have been kept secluded, apparently from time immemorial, in the parks of some of our aristocracy. [D] It has been, and still is, [Pg 141] a matter of lordly pride to their noble owners, that these cattle are held to be of a distinct and untameable race.

Feeling a full share of the interest attached to them, and anxious to gain the most accurate and circumstantial information, I was induced to pay a visit, during the summer of 1845, to the beautifully wooded and undulating Park of Chillingham, in which a herd of these cattle is preserved; and, although I have not been able to gather material for a perfect history of these animals, I think it will not be difficult to show that matters respecting them have been set forth as facts which are fictions; and that from some points of their history which have been correctly detailed, inferences have been drawn, which are by no means warranted by the facts.

In endeavouring to point out these errors and false reasonings, it will be necessary to make quotations from the old history of the white cattle, in Culley's 'Observations on Live Stock,' which has been so often repeated in works on natural history, and is, moreover, so thoroughly accredited, that it may now appear something like presumption to call it in question. To what extent it is called in question on the present occasion, and the reasons for so doing, will be seen in the running commentary which accompanies these quotations.

Culley says: "The Wild Breed, from being untameable, [Pg 142] can only be kept within walls or good fences; consequently very few of them are now to be met with, except in the parks of some gentlemen, who keep them for ornament, and as a curiosity: those I have seen are at Chillingham Castle, in Northumberland, a seat belonging to the Earl of Tankerville."

The statement of their being untameable is a mere assertion, founded upon no evidence whatever. But so far is it from being the fact, that, notwithstanding every means are used to preserve their wildness, such as allowing them to range in an extensive park — seldom intruding upon them — hunting and shooting them now and then — notwithstanding these means are taken to preserve their wildness, they are even now so far domesticated as voluntarily to present themselves every winter, at a place prepared for them, for the purpose of being fed. From which it may reasonably be concluded, that were they restricted in their pasture, gradually familiarised with the presence of human beings, and in every other respect treated as ordinary cattle, they would, in the course of two or three generations, be equally tame and tractable.

Whilst writing the foregoing I was not aware that any attempt had been made to domesticate these so-called untameable oxen; but on reading an account of these cattle by Mr. Hindmarsh, of Newcastle-upon-Tyne, (bearing date about 1837,) I find the following paragraph.

"By taking the calves at a very early age, and treating them gently, the present keeper succeeded in domesticating an ox and a cow.

They became as tame as domestic animals, and the ox fed as rapidly as a short-horned steer. He lived eighteen years, and when at his best was computed at 8 cwt. 14 lbs. The cow only lived five or six years. She [Pg 143] gave little milk, but the quality was rich. She was crossed by a country bull, but her progeny very closely resembled herself, being entirely white, excepting the ears, which were brown, and the legs, which were mottled." These facts speak for themselves.

Culley, in giving their distinguishing characteristics, says: "Their colour is invariably of a creamy white; muzzle black; the whole of the inside of the ear, and about one third of the outside, from the tips downwards, red; horns white, with black tips, very fine, and bent upwards; some of the bulls have a thin upright mane, about an inch and a half, or two inches long."

That their colour is invariably white is simply owing to the care that is taken to destroy all the calves that are born of a different description. It is pretty well known to the farmers about Chillingham (although pains are taken to conceal the fact,) that the wild cows in the park not unfrequently drop calves variously spotted. With respect to the redness of the ears, this is by no means an invariable character, many young ones having been produced without that distinctive mark; and Bewick records, that about twenty years before he wrote, there existed a few in the herd with *black* ears, but they were destroyed. So far from the character here given of the horns being confined to those white cattle, it is precisely the description of the horns of the Kyloe oxen, or black cattle. The investiture of some of the bulls with a mane is equally gratuitous; Cole, who was park-keeper for more than forty years, and of course had ample means of observation, distinctly informed me that they had no mane, but only some curly hair, about the neck, which is likewise an attribute of the Kyloe Oxen. [Pg 144]

Culley goes on to say: "From the nature of their pasture, and the frequent agitation they are put into by the curiosity of strangers, it is scarce to be expected that they should get very fat; yet the six years old oxen are generally very good beef, from whence it may be fairly supposed, that in proper situations they would feed well."

It would naturally be inferred from this, that the park in which they are kept is visited by strangers every day, who are allowed to drive them about, and disturb them in their feeding and ruminating, as boys hunt geese or donkeys on a common. This, however, is so far from being the case, that it frequently happens that the park is not visited for many weeks in succession, and certainly on an average it is not visited once a week. What is here meant by "the nature of their pasture," and "in proper situations they would feed well," it is difficult to say. The fact is, their pasture is both good and extensive, and they feed as well as animals always do who are left to themselves with plenty of food.

Their behaviour to strangers is thus described: "At the first appearance of any person, they set off at full speed, and gallop a considerable distance, when they make a wheel round, and come boldly up again, tossing their heads in a menacing manner; on a sudden, they make a full stop, at a distance of forty or fifty yards, looking wildly at the object of their surprise; but upon the least motion being made, they turn round again, and gallop off with equal speed; but forming a shorter circle, and, returning with a bolder and more threatening aspect, they approach much nearer, when they make another stand, and again gallop off. This they do several times, shortening [Pg 145] their distance, and approaching nearer, till they come within a few yards, when most people think it prudent to leave them."

In the instance in which I had an opportunity of witnessing their method of receiving visitors, the fashion was somewhat different. The park-keeper who accompanied me described, as we rode through the park in quest of them, what would be their mode of procedure on our approach. This he did from observations so repeatedly made, as to warrant him in saying that it was their invariable mode. It was perfectly simple, and I found it precisely as he had described it. When we came in sight of them, they were tranquilly ruminating under a clump of shady trees, some of the herd standing, others lying. On their first observing us, those that were lying rose up, and they all then began to move *slowly* away, not exactly to a greater distance from us, but in the direction of a thickly wooded part of the park, which was as distant on our left as the herd was on our right. To reach this wooded part they had to pass over some elevated ground. They continued to walk at a gradually accelerating pace, till they gained the most elevated part, when they broke out into a trot, then into a canter, which at last gave way to a full gallop, a sort of "devil-take-the-hindmost" race, by which they speedily buried themselves in the thickest recesses of the wood. What they may have done in Mr. Culley's time, we must take upon that gentleman's word; but at present, and for so long as the present park-keeper can recollect, they have never been in the habit of describing those curious concentric circles of which Mr. Culley makes mention in the last quotation.

The late mode of killing them is described as "perhaps [Pg 146] the only modern remains of the grandeur of ancient hunting. On notice being given, that a wild bull would be killed on a certain day, the inhabitants of the neighbourhood came mounted and armed with guns, &c., sometimes to the amount of a hundred horse, and four or five hundred foot, who stood upon walls or got into trees, while the horsemen rode off the bull from the rest of the herd until he stood at bay, when a marksman dismounted and shot. At some of these huntings twenty or thirty shots have been fired before he was subdued. On these occasions the bleeding victim grew desperately furious, from the smarting of his wounds, and the shouts of savage joy that were echoing from every side. But from the number of accidents that happened, this dangerous mode has been little practised of late years, the park-keeper alone generally shooting them with a rifled gun at one shot."

This vivid portraiture of a scene, which the writer is pleased to consider *grand*, does not appear to have much relation to the history of the *Genus Bos*: it however, exhibits the brutal and ferocious habits of two varieties of *Genus Homo*, namely *No*bility and *Mo*bility—two varieties which, although distinguished by some external marks of difference, possess in common many questionable characteristics.

Culley proceeds:—"When the cows calve, they hide their calves for a week or ten days in some sequestered situation, and go and suckle them two or three times a day. If any person come near the calves, they clap their heads close to the ground, and lie like a hare in form, to hide themselves; *this is a proof of their native wildness*, and is corroborated by the following circumstance [Pg 147] that happened to Mr. Bailey, of Chillingham, who found a hidden calf, two days old, very lean and very weak. On stroking its head it got up, pawed two or three times like an old bull, bellowed very loud, stepped back a few steps, and bolted at his legs with all its force; it then began to paw again, bellowed, stepped back, and bolted as before; but knowing its intention, and stepping aside, it missed him, fell, and was so very weak that it could not rise, though it made several efforts. But it had done enough: the whole herd were alarmed, and, coming to its rescue, obliged him to retire; for the dams will allow no person to touch their calves without attacking them with impetuous ferocity."

It seems almost unnecessary to remind the reader that all animals are naturally wild; and that even those animals that have been the longest under the dominion of man, are born with a strong tendency to the wild state, to which they would immediately resort, if left to themselves: it appears, therefore, rather gratuitous to tell us that the natural *actions of young animals* (whose parents have been allowed to run wild), *are proofs of their native mildness*!

The concluding paragraph requires no observation:—"When a calf is intended to be castrated, the park-keeper marks the place where it is hid, and, when the herd are at a distance, takes an assistant with him on horseback; they tie a handkerchief round the calf s mouth, to prevent its bellowing, and then perform the operation in the usual way. When any one happens to be wounded, or is grown weak and feeble through age or sickness, the rest of the herd set upon it, and gore it to death." [Pg 148]

The following engraving exhibits the effects of castration on the curvature and length of the horns.

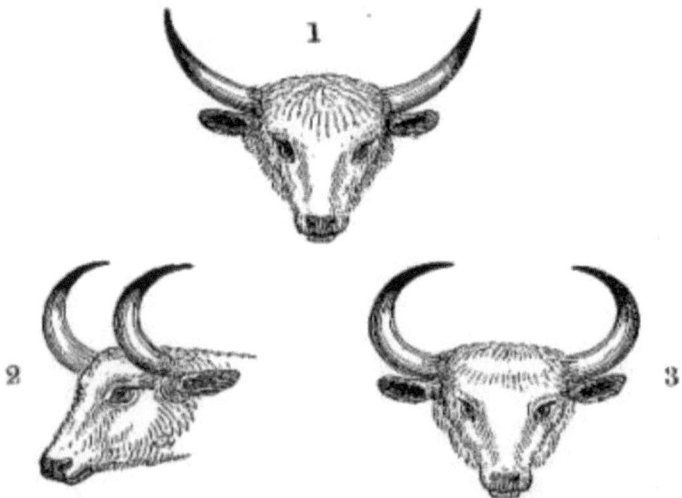

1. Head of the perfect animal. 2, 3. Heads of the emasculated animal.

We learn, on the authority of the present Lord Tankerville, that during the early part of the life-time of his father, the bulls in the herd had been reduced to three; two of them fought and killed each other, and the third was discovered to be impotent; so that the means of preserving the breed depended on the accident of some of the cows producing a bull calf.

In 1844 I wrote to Mr. Cole, the late park-keeper at Chillingham, requesting information on the following queries, to which he returned the answers annexed; and although they are not so explicit as might be wished, they embody facts both interesting and important. [Pg 149]

List of the Queries with their Answers.

1. How many pairs of ribs are there in the skeleton of the Chillingham Ox? *Thirteen pairs.*

2. How many vertebræ are there (from the skull to the end of the tail)? *Thirty in the back-bone, twenty in the tail.*

3. Will the wild cattle breed with the domestic cattle? *I have had two calves from a wild bull and common cow.*

4. What is the precise time the wild cow goes with young? *The same as the domestic cow.*

5. At what age does the curly hair appear which constitutes the mane of the wild bull? *They have no mane, but curly hair on their neck and head; more so in winter, when the hair is long.*

6. In what month does the rutting take place among the wild cattle? *At all times, – no particular time.*

J. Cole.

Here we have precise information on the following points:—namely, the number of ribs; the period of gestation; their having no mane; their not being in heat at any particular period; in all which points, they perfectly agree with the ordinary domestic cattle; and it is important to observe, that in the last point, namely, that of not being in heat at any particular time, they differ from every known *wild* species of cattle, among which the rutting season invariably occurs at a particular period of the year.

FOOTNOTES:

[D] Formerly these cattle were much more numerous, both in England and Scotland, than they are at present. Scanty herds are still preserved at the following places:—Chillingham Park, Northumberland; Wollaton, Nottinghamshire; Gisburne, in Craven, Yorkshire; Lime-hall, Cheshire; Chartley, Staffordshire; and Cadzow Forest, at Hamilton, Lanarkshire.

At Gisburne they are perfectly white, except the inside of their ears, which are *brown*.

From Garner's 'Natural History of Staffordshire,' we learn that the Wild Ox formerly roamed over Needwood Forest, and in the thirteenth century, William de Farrarus caused the park of Chartley to be separated from the forest, and the turf of this extensive enclosure still remains almost in its primitive state. Here a herd of wild cattle has been preserved down to the present day, and they retain their wild characteristics like those at Chillingham. They are cream-coloured, with *black muzzles and ears*; their fine sharp horns are also tipped with black. They are not easily approached, but are harmless, unless molested.

[Pg 150]

THE KYLOE, OR HIGHLAND OX.

Bos Taurus.

The Chillingham Cattle are *white*, and the Highland Cattle or Kyloes are generally *black*; but with this exception the same description might almost serve for both breeds.

In their natural and unimproved state, the Highland cattle are frequently well formed; their fine eyes, acute face, and lively countenances, give them an air of fierceness, which is heightened by their white, tapering, black-tipped, and sharp horns.

The Kyloe Oxen are very small (another respect in which they resemble the Chillingham Oxen). They likewise partake much of the nature of wild animals, which might be expected from the almost unlimited extent of their pasture, and their being but little subject to artificial treatment. [Pg 151]

Upon a close comparison of these two breeds, there appears not to be so much difference between the Highland cattle and the cattle of Chillingham as there is between any two breeds or varieties of

British cattle. Indeed so great is the similarity, that the Kyloe appears to be only a black variety of the Chillingham Ox, and the Chillingham Ox only a white variety of the Kyloe.

Dr. Anderson speaks of having seen a kind of Highland cattle which had a mane on the top of the head, of considerable length, and a tuft between the horns that nearly covered the eyes, giving them a fierce and savage aspect. He likewise mentions another kind which have hair of a pale lead colour, very beautiful in its appearance, and in its quality as glossy and soft as silk.

The Kyloe Oxen are natives of the Western Highlands and Isles, and are commonly called the Argyleshire breed, or the breed of the Isle of Skie, one of the islands attached to the county of Argyle. They are generally of a dark brown colour, or black, though sometimes brindled.

The Cows of the Isle of Skie (as is recorded by Martin, in his 'Description of the Western Islands of Scotland,') are exposed to the rigour of the coldest seasons, and become mere skeletons in the spring, many of them not being able to rise from the ground without help; but they recover as the season becomes more favorable, and the grass grows up; then they acquire new beef, which is both sweet and tender; the fat and lean is not so much separated in them as in other cows, but as it were larded, which renders it very agreeable to the taste. A cow in this isle may be twelve years old, when at the same time its beef is not above four, five, or six months old.

[Pg 152]

TABLE OF THE NUMBER OF VERTEBRÆ IN THE VARIOUS SPECIES OF THE GENUS BOS.

	Cerv.	Dors.	Lumb.	Sacr.	Caud.	Total.
American Bison	7	14	5	5	12+	
European Bison, or Aurochs	7	14	5	5	19	50
Yak	7	14	5	5	14	45

Gayal (Domestic)	7	14	5	5	16	47
Gayal (Asseel).						
Gyall						
Jungli Gau						
Italian Buffalo.						
Indian Buffalo.						
Skeleton of Buffalo in Surg. Coll. (locality unknown)	7	13	6	5	16	47
Gaur	7	13	6	5	19	50
Domestic Ox	7	13	6	5	21	52
Condore Buffalo						
Manilla Buffalo	7	13	6			
Pegasse						
Arnee						
Cape Buffalo	7	13	6	4	19	49
Zamouse (*Bos Brachyceros*)	7	13	6	4	20	50
Banteng of Java (*Bos Bantinger*)	7	13	6	4	18	48
Zebu, or Brahmin Ox	7	13	6	4	18	48
Galla Ox.						
Backeley (*Caffraria*).						
Musk Ox						

The osteological details in the above Table (except those of the Yak, which are given on the authority of Pallas) are from the Author's own observations.

TABLE OF THE PERIODS OF GESTATION OF THE VARIOUS SPECIES OF THE GENUS BOS.

	Periods
American Bison.	270 days. — Zool. Proc., 1849.
European Bison.	Between 9 and 10 months.
Gayal (Domestic)	Over 10 months
Gyall	11 months
Indian Buffalo	10 months 10 days.
Gaur	12 months
Domestic Ox.	270 days
Manilla Buffalo.	340 days
Arnee	12 months
Cape Buffalo	12 months
Zebu, or Brahmin Cow	300 days

Musk Ox	9 months

To supply the deficiencies in the foregoing Tables, the results of original observations are respectfully solicited. Address the Author or Publisher.

[Pg 154]

NOTE ON THE AMERICAN BISON.

It was Cuvier, I believe, who first made the statement, that the American Bison is furnished with *fifteen* pairs of ribs. In this particular he has been implicitly followed by every subsequent writer on the subject. Not being able to refer to a skeleton, and, moreover, never suspecting any inaccuracy in the statement, I followed the received account. But since this work has gone to press, I have had the opportunity of examining two skeletons, by which I find that—

The American Bison has only fourteen *pairs of ribs.*

I have, therefore, in the "Table of the Number of Vertebræ," (see p. 152,) set this species down as possessing only that number.

Of the two skeletons referred to (both of which are now in the British Museum), one is from a female Bison, some years a living resident in the Zoological Gardens; and the other is from a male, late in the possession of the Earl of Derby, at Knowsley, in Lancashire.

A corroborative circumstance (amounting, indeed, to a complete proof of the accuracy of these observations,) is presented by the fact, that, in both the cases *the number of lumbar vertebræ is precisely* five; thus making the true vertebræ to consist of nineteen, which Professor Owen [E] has shown to be the invariable number possessed by all ruminants.

FOOTNOTES:

[E] See, in the Proceedings of the Zoological Society, Professor Owen's 'Account of his Dissection of the Aurochs.'

[Pg 155]

APPENDIX

THE FREE MARTIN.

Cows usually bring forth but one calf at a birth; occasionally, however, they produce twins. John Hunter, in his 'Observations on the Animal Economy,' says: "It is a fact known, and I believe almost universally understood, that when a cow brings forth two calves, one of them a bull-calf, and the other to appearance a cow, that the cow-calf is unfit for propagation; but the bull-calf grows up into a very proper bull. Such a cow-calf is called, in this country, a Free Martin, and is commonly as well known among the farmers as either cow or bull. It has all the external marks of a cow-calf, namely, the teats, and the external female parts, called by farmers the bearing. It does not show the least inclination for the bull, nor does the bull ever take the least notice of it. In form it very much resembles the Ox, or spayed heifer, being considerably larger than either the bull or the cow, having the horns very similar to the horns of an Ox. The bellow of the Free Martin is similar to that of an Ox, having more resemblance to that of the cow than that of the bull."

Free Martins are very much disposed to grow fat with good food. The flesh, like that of the Ox or spayed heifer, is generally much finer in the fibre than either the bull or cow; is even supposed to exceed that of the Ox and heifer [Pg 156] in delicacy of flavour, and bears a higher price at market. However this superiority of the flavour does not appear to be universal, for Mr. Hunter was informed of a case which occurred in Berkshire, in which the flesh of a Free Martin turned out nearly as bad as bull beef. This circumstance probably arose from the animal having more the properties of a bull than a cow.

Mr. Hunter, having had many opportunities of dissecting Free Martins, has satisfactorily shown that their incapacity to breed, and all their other peculiarities, result from their having the generative organs of both sexes combined, in a more or less imperfect state of development, in some cases the organs of the male preponderating, in others those of the female.

The above, which is copied from an engraving in Hunter's work on the 'Animal Economy,' is a representation of a Free Martin, five years old; it shows the external [Pg 157] form of that animal, which is neither like the bull nor cow, but resembling the Ox or spayed heifer.

Although, as Hunter observes, "it is almost universally understood, that when a cow brings forth two calves, one of them a bull-calf, and the other to appearance a cow, that the cow-calf is unfit for propagation," it is by no means universally the fact, as instances of such twins breeding were known even in Hunter's time, and have been witnessed more recently. The following is recorded in Loudon's 'Mag. of Nat. History,' and occurred a few years previous to 1826: Jos. Holroyd, of Withers, near Leeds, had a cow which calved twins, a bull-calf and a cow-calf. As popular opinion was against the cow-calf breeding, it being considered a Free Martin, Mr. Holroyd was determined to make an experiment of them, and reared them together. They copulated, and in due time the heifer brought forth a bull-calf, and she regularly had calves for six or seven years afterwards.

"If," says Hunter, "there are such deviations as of twins being perfect male and female, why should there not be, on the other hand, an hermaphrodite, produced singly, as in other animals? I had the examination of one which seemed, upon the strictest inquiry, to have been a single calf; and I am the more inclined to think this true, from having found a number of hermaphrodites among black cattle, without the circumstance of their birth being ascertained."

If Hunter had carried this reasoning a little further, he might have asked,—Why should there not be a Free Martin, or hermaphrodite, produced in the case of twins, when they are both apparently males, or both apparently females? Had he done this, he would not, probably, [Pg 158] have made the following observation: "I need hardly observe, that if a cow has twins, and they are both bull-calves, they are in every respect perfect bulls; or if they are both cow-calves, they are perfect cows." What is this but saying that a bull-calf is a bull-calf, and a cow-calf is a cow-calf? For a Free Martin, or hermaphrodite, is not, in any case, either a bull or a cow.

There does not appear to be anything known of the peculiar circumstances under which, what is termed a Free Martin is produced.

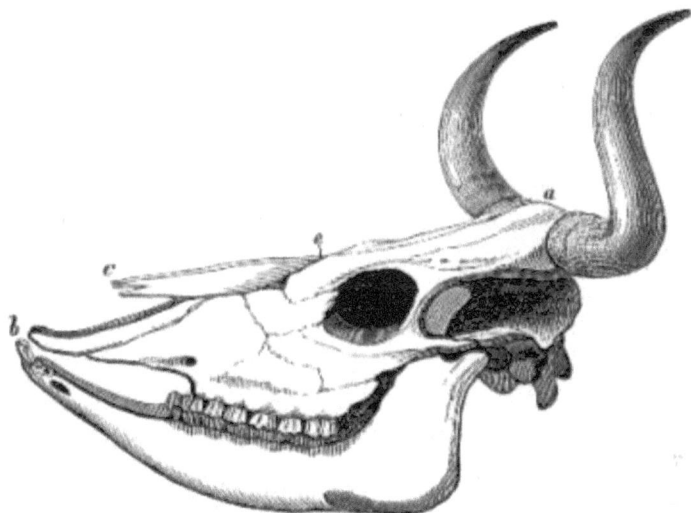

Skull of Domestic Ox.

The most general observation that can be made on the subject appears to be, that cows sometimes produce calves, which, by reason of their imperfectly developed generative system, are incapable of procreating. [Pg 159]

THE SHORT-NOSED OX.

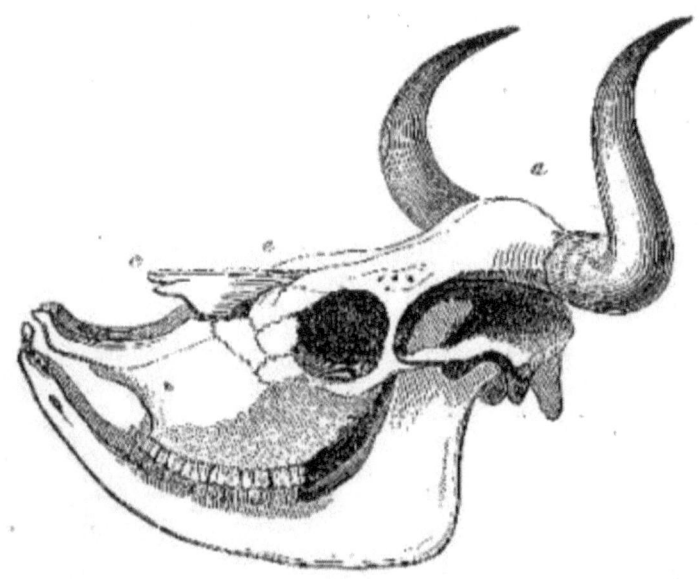

Skull of short-nosed Ox of the Pampas.

The common Ox, originally taken over to America by the early Spanish settlers, now runs wild in immense herds on the Pampas, where it is hunted and slain for its hide. Some idea may be formed of the immensity of these herds, from the circumstance that nearly a million of hides are annually exported from Buenos Ayres and Monte Video to Europe.

Some of the herds in these wild regions have undergone a most singular modification of the cranium, consisting in a shortening of the nasal bones, together with the superior and inferior maxillaries. There is a skull of this variety in the Museum of the College of Surgeons, of which the above is a sketch. [Pg 160]

ON THE UTILITY OF THE OX TRIBE TO MANKIND.

How eminently serviceable to man these animals are, is shown in the following table, in which are set forth the most important uses to which their various parts are applied:

Skin.—The skin has been of great use in all ages. The ancient Britons constructed their boats with osiers, and covered them with the hides of bulls; and these boats were sufficiently strong to serve for short coasting voyages. Similar vessels are still in use on the Irish lakes, and in Wales on the rivers Dee and Severn. In Ireland they are called *curach*, in England *coracles*, from the British *cwrwgl*, a word signifying a boat of that structure.

Boots, shoes, harness, &c. for horses, and various kinds of travelling trunks are made from hides when tanned. The skin of the calf is extensively used in the binding of books, and the thinnest of the calf skins are manufactured into vellum. The skin of the Cape Buffalo is made into shields and targets, and is so hard that a musket ball will scarcely penetrate it.

Hair.—The short hair is used to stuff saddles and other articles; also by bricklayers in the mixing up of certain kinds of mortar. It is likewise frequently used in the manuring of land. The *long* hair from the tail is used for stuffing chairs and cushions. The hair of the Bison is spun into gloves, stockings, and garters, which are very strong, and look as well as those made of the finest sheep's wool; very beautiful cloth has likewise been manufactured from it. The Esquimaux convert the skin [Pg 161] covering the tail into caps, which are so contrived that the long hair falling over their faces, defends them from the bites of the mosquitoes.

Horns.—The horns of cattle consist of an outside horny case, and an inside conical-shaped substance, somewhat between hardened hair and bone. The horny outside furnishes the material for the manufacture of a variety of useful articles. The first process consists in cutting the horn transversely into three portions.

1. The *lowest* of these, next the root of the horn, after undergoing several operations by which it is rendered flat, is made into combs.

2. The *middle* of the horn, after being flattened by heat, and its transparency improved by oil, is split into thin layers, and forms a

substitute for glass in lanterns of the commonest kind. [The merit of the invention of these horn plates, and of their application to lanterns, is ascribed to King Alfred, who is said to have first used lanterns of this description to preserve his candle time-measurers from the wind.]

3. The *tips* of the horns are generally used to make knife-handles; the largest and best are used for crutch-stick heads, umbrella handles, and ink-horns, and the smallest and commonest serve for the tops and bottoms of ink-horns.

Spoons, small boxes, powder flasks, spectacle frames, and drinking horns are likewise made of the outer horny case.

The interior or core of the horn is boiled down in water, when a large quantity of fat rises to the surface; this is sold to the makers of yellow soap. — The liquid itself is used as a kind of glue, and is purchased by the cloth-dressers for stiffening. — The bony substance which remains [Pg 162] behind, is ground down, and sold to the farmers for manure.

Besides these various purposes to which the different parts of the horn are applied, the chippings which arise in comb-making are sold to the farmer for manure, at about one shilling a bushel. In the first year after they are spread over the soil they have comparatively little effect; but during the next four or five their efficiency is considerable. The shavings, which form the refuse of the lantern-maker, are of a much thinner texture. Some of them are cut into various figures, and painted and used as toys; for they curl up when placed in the palm of a warm hand. But the greater part of these shavings are sold also for manure, which from their extremely thin and divided form, produce their full effect upon the first crop.

Feet. — An oil is extracted from the feet of oxen — hence called Neat's-foot-oil — of great use in preparing and softening leather.

Skin, *horns*, *hoofs*, and *cartilages* are used to make glue.

Blood is used in the formation of mastic; also in the refining of sugar, oil, &c.; and is an excellent manure for fruit trees.

Blood, *horns*, and *hoofs* in the formation of Prussian blue.

Gall is used to cleanse woollen garments, and to obliterate greasy and other stains.

Suet, Fat, Tallow are chiefly manufactured into candles; they are also used to precipitate the salt that is drawn from briny springs.

Intestines, when dried, are used as envelopes for German and Bologna sausages; in some countries to [Pg 163] carry butter to market. By gold-beaters, in the process of making gold-leaf. Gold-beater's skin, as it is called, forms the most innocent sticking plaster for small cuts on the hands or fingers.

The Stomachs vulgarly called *inwards*, after being washed and boiled, are sold as an article of food under the name of *tripe*.

The Excrementitious matters are used to manure the land.

The Bones are used as a substitute for ivory in the manufacture of a variety of small articles of a common kind; also for manuring land. "When calcined they are used as an absorbent to carry off the baser metals in refining silver. From the tibia and carpus is procured an oil much used by coach-makers and others in dressing and cleaning harness, and all trappings belonging to carriages."

Flesh, both fresh and salted, is generally esteemed as an article of food. *Pemmican* is made of the flesh of the American Bison: this is dried in the sun by the Indians, spread on a skin, and pounded with stones. When the Indians have got it into this state, they sell it to the different forts, where all the hair is carefully sifted out of it, and melted fat kneaded into it. If it be well made, and kept dry, it will not spoil for a year or two.

Milk, a nutritious beverage, *per se*, is used in the composition of innumerable articles of diet; from milk is obtained cream, butter, and cheese. [Pg 164]

SOME ACCOUNT OF THE ALPINE COWHERDS,

WITH A NOTICE OF THE CELEBRATED SWISS AIR

The Ranz des Vaches.

In the Alps, fine cattle are the pride of their keeper, who, not being satisfied with their natural beauty, also gratifies his vanity by adorning his best cows with large bells, suspended from broad thongs. Every *Senn*, or great cow-keeper, has a harmonious set of bells, of at least two or three, chiming in accordance with the famous *Ranz des Vaches*. The finest black cow is adorned with the largest bell, and those next in appearance wear the two smaller ones.

It is only on particular occasions that these ornaments are worn, namely, in spring, when they are driven to the Alps, or removed from one pasture to another; or in their autumnal descents, when they travel to the different farmers for the winter. On such days the Senn, even in the depth of winter, appears dressed in a fine white shirt, with the sleeves rolled above the elbows; neatly embroidered red braces suspend his yellow linen trowsers, which reach down to the shoes; he wears a small leather cap on his head, and a new and skilfully carved wooden milk-bowl hangs across his left shoulder. Thus arrayed, the Senn proceeds, singing the *Ranz des Vaches*, followed by three or four fine goats; next comes the finest cow, adorned with the great bell; then the other two with the smaller bells; and these are succeeded by the rest of the cattle, walking one after another, and having in their rear the bull, with a one-legged milking-stool on his horns; the procession is closed by a *traineau*, or sledge, bearing the dairy implements. [Pg 165]

When dispersed on the Alps, the cattle are collected together by the voice of the Senn, who is then said to allure them. How well these cows distinguish the voice of their keeper, appears from the circumstance of their hastening to him, although at a great distance, whenever he commences singing the *Ranz des Vaches*.

This celebrated air is played on the bagpipes, as well as sung by the young Swiss cowherds while watching their cattle on the mountains. The astonishing effects of this simple melody on the Swiss soldier, when absent from his native land, are thus described by Rousseau:

"Cet air, se chéri des Suisses qu'il fut défendu sous peine de mort de le jouer dans leurs troupes, parce qu'il faisait fondre en larmes, déserter, ou mourir, ceux qui l'entendaient, tant il excitait en eux l'ardent desir de revoir leur pays. On chercherait en vain dans cet air les accens énergetiques capables de produire de si étonnans effets. Ces effets, qui n'ont aucun lieu sur les étrangers, ne viennent qui de l'habitude, des souvenirs de mille circonstances qui, retracées par cet air à ceux que l'entendent, et leur rappellant leur pays, leurs anciens plaisirs, leur jeunesse, et toutes leur façons de vivre, excitent en eux une douleur amère d'avoir perdu tout cela. La musique alors n'agit point précisément comme musique, mais comme signe memoratif. Cet air, quoique toujours le même, ne produit plus aujourd'hui les mêmes effets qu'il produisait ci-devant sur les Suisses, parce qu'ayant perdu le gôut de leur première simplicité, ils ne la regrettent plus quand on la leur rappelle. Tant il est vrai que ce n'est pas dans leur action physique qu'il faut chercher les plus grand effets des sons sur le cœur humain." [Pg 166]

For the delectation of the musical reader, the notes of this celebrated air are here introduced, with the words, and an English imitation:

[Pg 167]

The words are as follows:—

>Quand reverai-je en un jour,
>Tous les objets de mon amour,
>Nos clairs ruisseaux,
>Nos hameaux,
>Nos côteaux,
>Nos montagnes,
>Et l'ornament de nos montagnes,

La si gentille Isabeau?
Dans l'ombre d'un ormeau,
Quand danserai-je au son du Chalameau?
Quand reverai-je en un jour,
Tous les objets de mon amour,
Mon père,
Ma mère,
Mon frère,
Ma soeur,
Mes agneaux,
Mes troupeaux,
Ma bergère?

IMITATED.

When shall I return to the Land of the Mountains—
The lakes and the Rhone that is lost in the earth—
Our sweet little hamlets, our villages, fountains,
The flour-clad rocks of the place of my birth?
O when shall I see my old garden of flowers,
Dear Emma, the sweetest of blooms in the glade,
And the rich chestnut grove, where we pass'd the long hours
With tabor and pipe, while we danced in the shade?
When shall I revisit the land of the mountains,
Where all the fond objects of memory meet:
The cows that would follow my voice to the fountains,
The lambs that I called to the shady retreat:
My father, my mother, my sister, and brother;
My all that was dear in this valley of tears;
My palfrey grown old, but there's ne'er such another;
My dear dog, still faithful, tho' stricken in years:
The vesper bell tolling, the loud thunder rolling,
The bees that humm'd round the tall vine-mantled tree:
The smooth water's margin whereon we were strolling
When evening painted its mirror for me?
And shall I return to this scenery never?
These objects of infantine glory and love,—
O tell me, my dear Guardian Angel, that ever
Floats nigh me,—safe guide to the regions above.
[Pg 168]

SYNOPTICAL TABLE OF HABITAT

Buffalo—*Bos Bubalus*	Asia, North Africa, and South Europe.
Manilla Buffalo	Island of Manilla.
Condore Buffalo	Island of Pulo Condore.
Cape Buffalo	South Africa.
Pegasse	Congo, Angola, Central Africa.

Arnee	India and China.
Gaur	India.
American Bison	North America.
Aurochs	Lithuania.
Yak	Tartary and Hindustan.
Musk Ox	North America.
Zamouse, or Bush Cow	Gambia, Sierra Leone.
Banteng	Island of Java.
Gyall	India.
Gayal	India.
Sanga, or Galla Ox	Abyssinia.
Zebu — Brahmin Ox	Southern Asia, Eastern Africa.
Domestic Ox	Generally diffused.

[Pg 169]

AND MODE OF LIFE.

Mode of Life.

Partial to water and mud, swampy localities.

Semi-aquatic in its habits, — sometimes called the Water Buffalo.

Fond of wallowing in mire, and swims well.

Lives much in the water, and feeds on aquatic plants.

Ranges in mountain forests, and feeds on leaves and buds of trees.

Migratory in its habits—fond of bathing in marshy swamps.

Lives chiefly on the woody banks of rivers—feeds on bark of trees, lichens, and herbaceous plants.

Feeds on the short herbage peculiar to the tops of mountains and bleak plains.

Lives chiefly on rocky mountains.

Delights in the deepest jungles—feeds on leaves and shoots of brushwood.

Lives entirely on woody-mountains—feeds on shoots and shrubs.

Half domesticated.

Domesticated, and artificially fed.

So completely domesticated, as to be subject to an endless variety of diseases, and generally requires medical attendance.

[Pg 170]

THE INDEFINITE DEFINITIONS OF COL. HAMILTON SMITH.

On commencing this Monograph of the *Genus Bos*, I entertained the confident expectation, that in the voluminous work of Cuvier's 'Animal Kingdom,' translated and enlarged by Griffith and others, I should find all that related to generic and specific distinction so clearly exhibited, and so systematically arranged, that I should have no hesitation in adopting the classification there set forth, and no difficulty in determining the place of any new species or variety.

With this expectation I diligently studied that portion of Col. H. Smith's volume on the Ruminantia, which treat of the *Genus Bos*, and I here subjoin (verbatim) the generic and subgeneric characters there given of that Genus, by which it will be seen how far they fall short of the clearness and precision which are indispensable to a scientific work.

GENERIC CHARACTERS.

"*Genus BOS.*—Skull very strong, dense about the frontals, which are convex, nearly flat, or concave; horns invariably occupying the crest, projecting at first laterally; osseous nucleus throughout porous, even cellular; muzzle *invariably broad*, naked, moist, *black*; ears, *in general, middle sized*; body *long*; legs *solid*; stature *large*."

Generic characters should be such as will apply to every species in the genus; they should likewise be such as will [Pg 171] distinguish the genus described from every other genus. From such observations as I have been enabled to make, the five last-mentioned characters do not appear to accord with either of these conditions.

1st. The muzzle is stated to be *black*; but in the Yak, and in domestic cattle (as may be observed by any one), the muzzle is very frequently *white*; and granting that it was invariably black, other genera of the ruminantia have the muzzle black: and therefore it cannot be said to be a distinguishing mark of the *Genus Bos*.

2d. The ears are stated to be *in general middle-sized*. To pass over the extreme vagueness of the terms "*in general*" and "*middle-sized*," I may state that having measured the ears of several species, I find them to be of all lengths, varying from 5 inches to nearly 18 inches. Such a term as "*middle-sized*" may be applied "*in general*" to the ears of a vast variety of animals; and therefore it cannot be applied *in particular* to the *Genus Bos*.

3d. The body is said to be *long*. They are, indeed, of all lengths, from 4 ft. 6 in. to nearly 11 ft. Can the term long be equally applicable to animals of such different lengths?

4th. The legs are said to be *solid*. In some species the legs are very slender, as the Zebu, Manilla Buffalo, and Domestic Ox.

5th. The stature is said to be *large*. From actual measurement I find the stature to vary from 2 ft. 8 in. to upwards of 6 ft.; the smaller species weighing not more than 100 lbs., the larger weighing as much as 2000 lbs. Can the term large be equally applicable to animals of such different sizes? [Pg 172]

SUB-GENERIC CHARACTERS.

"*Sub-genus* I.—*Bubalus*.—Animals low in proportion to their bulk; limbs very solid; head large, forehead narrow, very strong, convex; chaffron straight; muzzle square, horns lying flat, or bending laterally with a certain direction to the rear; eyes large; ears mostly funnel-shaped; no hunch; a small dewlap; *female udder with four mammæ; tail long*; slender."

This sub-genus comprises Cape Buffalo, Pegasse, Arnee, Domestic Buffalo.

"*Sub-genus* II.—Bison.—Forehead slightly arched, much broader than high; horns placed before the salient line of the frontal crest; the plane of the occiput forming an obtuse angle with the forehead and semicircular in shape; fourteen or fifteen pairs of ribs; the shoulders rather elevated; the *tail shorter*; the legs more slender; the tongue blue; the hair soft and woolly."

This sub-genus comprises Aurochs, Gaur, American Bison, Yak, Gayal.

"*Sub-genus* III.—Taurus.—Forehead square from the orbits to the occipital crest, somewhat concave, not convex, or arched as in the former; the horns rising from the sides of the salient edge or crest of the frontals; the plane of the occiput forming an acute angle with the frontal, and of quadrangular form; the curve of the horns outwards, upwards, and forwards; no mane; a deep dewlap; *thirteen pairs of ribs; tail long; udder four teats in a square.*"

This sub-genus comprises the Urus and the Domestic Ox. [Pg 173]

Subgeneric characters should be such as will clearly distinguish the animals of one sub-genus from those of another. But here we have set down, in the sub-genus Bubalus, tail *long*, slender; in the sub-genus Taurus, tail *long*; and although the epithet slender is not added in the latter case, yet in truth it ought to be, as the tail of Taurus is quite as slender as that of Bubalus.

The udder of Bubalus is said to have four mammæ; they are not stated to be in a square, but, on examination, I find they are so; the udder of Taurus has likewise four teats in a square.

Thirteen pairs of ribs are set down as a distinguishing character of the sub-genus Taurus; but the Cape Buffalo, Domestic Buffalo, and the Manilla Buffalo (in the sub-genus Bubalus), and the Gaur (in the sub-genus Bison), all possess thirteen pairs of ribs.

In the sub-genus Bison the tail is said to be *shorter* than the tail of Bubalus; but on subjecting them to the infallible test of feet and inches, I find the tails of the Aurochs, Gaur, Yak, and Gayal, to be decidedly *longer* than those of the Cape or the Manilla Buffalo.

The legs of Bisons are stated to be more slender than those of Buffaloes, — the reverse of this is the fact in the instances which I have had an opportunity of observing.

SPECIFIC DETAILS.

The details of a system of scientific classification should be precise, methodical, and consistent; but the method observed by Col. Smith, in describing the lengths of animals, can scarcely be called either precise or consistent; for example, he states: [Pg 174] —

1st. That the Cape Buffalo is nine feet from *nose to root of tail*.

2d. That the Gaur is twelve feet long *to the end of tail*.

3d. That the Aurochs is ten feet three inches *from nose to tail*.

4th. That the Domestic Buffalo is eight feet six inches long, *without mentioning either nose or tail*.

In none of these cases can we be even proximately certain of the length of the animal.

In the first instance we may err to the amount of the length of the head; as it is not stated whether the measure was taken when the head was extended in a line with the back, or in a position at right angles with the back, or in any intermediate position.

The following outline will illustrate this: —

It is obvious that the length of a line from the nose to the tail will vary according to the different positions of the head of the animal.

In the second instance (taking it for granted that the measure was taken from the nose), the same difficulty exists with respect to the head, and another difficulty presents itself in our being left to guess the length of the tail, which might be eighteen inches, or it might be four feet. [Pg 175]

In the third instance, the same difficulty exists with respect to the head, and the difficulty is further complicated by our being left to guess whether the root or the end of the tail is meant.

In the fourth we are completely "*at sea.*"

The true value of these characteristic distinctions, definitions, or descriptions, are left to the appreciation of the judicious reader. Colonel Smith may doubtless be, what he has been styled, "an indefatigable naturalist," and "in general" an exact one; but in this special instance of the *Genus Bos*, his warmest admirers must allow that his accuracy and precision have not kept pace with his industry.

Hungarian Ox, Bos Taurus, from a specimen in the British Museum.

[Pg 176]

MR. SWAINSON'S TRANSCENDENTAL ATTEMPT AT CLASSIFICATION.

The following very laboured attempt to arrange the various species of *Genus Bos* into groups, according to the Quinary or Circular System of M'Leay, is from the pen of Mr. Swainson—the precise and fastidious Swainson—who, from the number and boldness of his hypothetical views in every department of Zoology, may be truly regarded as the beau-ideal of a speculative naturalist—one of those, in short, so well described by Swift, "whose chief art in division hath been to grow fond of some proper mystical number, which their imaginations have rendered sacred to a degree, that they force common reason to find room for it in every part of nature; *reducing, including,* and *adjusting,* every *genus* and *species* within that compass, by coupling some against their wills, and banishing others at any rate."

After describing the various members of the Bovine Family according to the Procrustean method of stretching and chopping, Mr. Swainson continues in his peculiarly dogmatic style "The types of form of the *Genus Bos*, above enumerated, *we shall now demonstrate to be a natural group.* We have seen that the first represented by the *Bos Scoticus*, or Scotch Wild Ox, is an untameable savage race, which preserves, even in the domestication of a park, all that fierceness which the ancient writers attributed to the Wild Bulls of Britain and of the European Continent. Let those who imagine that the [Pg 177] influence of civilization, of care, and of judicious treatment, will alter the natural instincts of animals, look to this as a palpable refutation of their doctrine. Where is that boasted power of man over nature? Where the fruits of long-continued efforts and fostering protection? The *Bos Scoticus* is as untameable now as it was centuries ago, simply for this reason, that it is in accordance with an unalterable law of nature; a law by which one type in every circular group is to represent the worst passions of mankind—fierceness, or cruelty, or horror. In the *Urus* we consequently have the type of the wild and untameable *Feræ* among quadrupeds, the eagles among birds, and the innumerable analogies which all the subordinate groups of these two great divisions present. Following this is the typical Ox—a god among the ancients, and that animal above all others, which, from its vital importance to man, we should naturally expect such a nation as the ancient Egyptians would exalt above all others. It is, in short, the typical perfection of the whole order of Ruminants, and consequently represents the *Quadrumana* among quadrupeds, and the *Incessores* among birds. The third type is no less beautiful; but it cannot be illustrated without going into details which it is not our present intention to make public: suffice it, however, to say, that in the prominent hump upon the shoulders we have a perfect representation of the Camel, one of the most striking types of the order, while it reminds us at the same time of the Buffalo, the genus *Acronatus* among the large Antelopes, and numerous other representations of the same form. The fourth type is our *Bos Pusio*: here we find the horns, when present, remarkably small, but in many cases absent; and [Pg 178] the size is diminutive to an extreme. These also are distinguishing marks of the groups it is to represent: the *Tenuirostres* among birds, and the *Glires*, or mice, among quadrupeds, are the smallest of their respective classes; and both are typically distinguished by wanting all appendages to the head, either in the form of crests or horns. The fifth type is, perhaps, the most extraordinary of all; it should represent not only the order

Rasores among birds, but also the *Camelopardalis* among ruminating quadrupeds. Hence we find that, in accordance with the first of these analogies, it is a peaceful domesticated race, and that it has horns of an unusually large size, even in its own group; while, at the same time, those horns have that peculiar structure which can only be traced in the Camelopardalis; they are covered with skin, which passes so imperceptibly to the horny state, that, as Captain Clapperton observes, "there is no exact demarcation where the one commences and the other ends." The five leading types of quadrupeds and birds being now represented, and in precisely the same order, *we demonstrate* the groups to be natural by the following table: —

Genus BOS — *the Natural Types.*

1. *Bos Scoticus.*	Fierce, untameable.		Feræ.	Raptores.
2. — — *Taurus.*	Pre-eminently typical.		Primates.	Incessores.
3. — — *Dermaceros.*	Appendages on the head greatly developed		Ungulata.	Rasores.
4. — — *Pusio.*	Stature remarkably small.		Glires.	Grallatores.
5. — — *Thersites.*	Fore-part of the shoulders elevated		Cetacea.	Natatores.

In regard to the last type, the analogies can only be [Pg 179] traced through the animals or types of other groups; but should the habits of *Thersites* lead it to frequent the water (like the Buffaloes) more than any other species of true oxen — a supposition highly probable — the analogy to the *Cetacea* and the *Natatores* would be direct. When we find in all the other four types such a surprising representation of the same peculiarities, we are justified in believing that want of information alone prevents this analogy from being so complete as the others. These analogies, in point of fact, may be traced through the whole of the principal groups in this order, the most important, and the most numerous of ungulated animals." Our luminous classifier then triumphantly winds up: — "*Having now demonstrated*, in one of the very lowest groups of quadrupeds, the validity of those principles of natural classification we have so often illustrated," &c.

Let us not be confounded with high-sounding terms; let us rather endeavour to ascertain the meaning of them, if indeed they possess a meaning. Here we have, under the head of "*Genus* Bos — the Natural Types" — (see p. 178), certain words arranged in regular columns, which, at a first glance, appear as though they were intended to bear some relation to each other. But let us ask the most ordinary observer, or the most profound observer, or the observer of any grade or shade between these two extremes, what resemblance — what relation — what analogy — can be discovered between an ordinary bull (*Taurus*) and a man, a monkey, or a bat (*Primates*); or between Taurus and the *Incessores* (Perching Birds)? Or between Buffaloes, whose horns are partially covered with skin (*Dermaceros*), and cocks and hens (*Rasores*)? Can any one say wherein consists the similarity between a dwarf [Pg 180] Zebu and a Mouse, or a Flamingo? Yet this is the material of which the columns are composed.

But one of the most unhappy of Mr. Swainson's speculations is that wherein he represents the *Bos Scoticus*, or wild ox, as the type of "an *untameable savage* race, which preserves, even in the domestication of a park, all that fierceness which the ancient writers attributed to the wild bulls of Britain and the European continent. Let those who imagine that the influence of civilization, of care, and of judicious treatment, will alter the natural instinct of animals, look to this as a palpable refutation of their doctrine. [!] Where is that boasted power of man over nature? Where the fruits of long-continued efforts and fostering protection? [!!] The *Bos Scoticus* is as untameable now as it was centuries ago, simply for this reason, that it is in accordance with an unalterable law of nature; a law by which one type in every group is to represent the worst passions of mankind — fierceness, or cruelty, or horror." [!!!]

Who would for a moment imagine that all this grandiloquence is bestowed upon an animal, which is so far from being fierce and untameable, that young ones, taken and reared with ordinary cattle, become, even in the first generation, as tame as domestic animals? [See account of Chillingham White Cattle, p. 140.]

For a more complete satisfaction of his thought, the reader is referred to Mr. Swainson's volume "On the Natural History and Classification of Quadrupeds," p. 274, where he has given us an incoherent abstract of Colonel Smith's article on the *Bovinæ*, without, however, making the least attempt to verify the statements there recorded. The descriptions and characteristics are [Pg 181] avowedly Colonel Smith's; but, in justice to the latter gentleman, it must be added, that the disquisitions on the circular succession of forms, and the analogical relations, are entirely Mr. Swainson's.

ON SPECIES AND VARIETY.

What constitutes a species? And how far do the limits of varieties extend? Cuvier, who is, perhaps, the best authority we can have upon this subject, in defining a species, says: — *A species comprehends all the individuals which descend from each other or from a common parentage, and those which resemble them as much as they do each other.* Thus, the different races which they have generated from them are considered as varieties but of one species. Our observations, therefore, respecting the differences between the ancestors and the descendants, are the only rules by which we can judge on this subject; all other considerations being merely hypothetical, and destitute of proof. Taking the word *variety* in this limited sense, we observe that the differences which constitute this variety depend upon determinate circumstances, and that their extent increases in proportion to the intensity of the circumstances which occasion them.

Upon these principles it is obvious, that the most superficial characters are the most variable. Thus colour depends much upon light; thickness of hair upon heat; [Pg 182] size upon abundance of food, &c. In wild animals, however, these varieties are greatly limited by the natural habits of the animal, which does not willingly migrate from the places where it finds, in sufficient quantity, what is necessary for the support of its species, and does not even extend its haunts to any great distances, unless it also finds all these circumstances conjoined. Thus, although the Wolf and the Fox inhabit all the climates from the torrid to the frigid zone, we hardly find any other differences among them, through the whole of that vast space, than a little more or less beauty in their furs. The more savage animals, especially the carnivorous, being confined within narrower limits, vary still less; and the only difference between the Hyæna of Persia and that of Morocco, consists in a thicker or a thinner mane.

Wild animals which subsist upon herbage, feel the influence of climate a little more extensively, because there is added to it the influence of food, both in regard to its abundance and its quality. Thus the Elephants of one forest are larger than those of another; their tusks also grow somewhat longer in places where their food may happen to be more favorable for the production of the substance of ivory. The same may take place in regard to the horns of Stags and Rein-deer. Besides, the species of herbivorous animals, in their wild state, seem more restrained from migrating and dispersing than the carnivorous species, being influenced both by climate, and by the kind of nourishment which they need.

We never see, in a wild state, intermediate productions between the Hare and the Rabbit, between the Stag and the Doe, or between the Martin and the Weasel. Human artifice contrives to produce all these [Pg 183] intermixtures of which the various species are susceptible, but which they would never produce if left to themselves.

The degrees of these variations are proportional to the intensity of the causes that produce them, namely, the slavery or subjection under which these animals are to man. They do not proceed far in half-domesticated species.

In the domesticated herbivorous quadrupeds, which man transports into all kinds of climates, and subjects to various kinds of management, both in regard to labour and nourishment, he procures certainly more considerable variations, but still they are all merely superficial: greater or less size; longer or shorter horns, or even the want of these entirely; a hump of fat, larger or smaller, on the shoulder; these form the chief differences among particular races of the *Bos Taurus*, or domestic Black Cattle; and these differences continue long in such breeds as have been transported to great distances from the countries in which they were originally produced, when proper care is taken to prevent crossing.

Nature appears also to have guarded against the alterations of species which might proceed from mixture of breeds, by influencing the various species of animals with mutual aversion. Hence all the cunning and all the force that man is able to exert is necessary to accomplish such unions, even between species that have the nearest resemblance. And when the mule-breeds that are thus produced by these forced conjunctions happen to be fruitful, which is seldom the case, this fecundity never continues beyond a few generations, and would not probably proceed so far, without a continuance of the same causes which excited it at first. [Pg 184]

This being the case, it is quite clear that the fact of two animals producing an intermediate race is no proof whatever of their specific identity; for it is well known, and has been already alluded to, that several animals. Birds as well as Mammalia, produce offspring, and are nevertheless distinct, both as it regards anatomical structure and external form.

Neither does it constitute the species identical if either or both the hybrids be even capable of fruitful intercourse with the original or parent species. Hamilton Smith goes so far as to say, that "if it even were proved that a prolific intermediate race exist, produced by the intermixture of both, it would not fully determine that both form only one original species: what forms a species, and what a variety, is as yet far from being well understood."

It is, however, pretty generally agreed, that animals are of the same species, that is to say, have been derived from one common stock, when their offspring have the power, *inter se*, of indefinitely continuing their kind; and conversely, that animals of distinct species, or descendants of stocks originally different, cannot produce a mixed race which shall possess the capability of perpetuating itself.

To conclude, it must be obvious, that permanent anatomical differences are the only true criteria of distinctions of species. [Pg 185]

THE BANTENG OF JAVA.

Bos Bantinger, or Bantiger. Bos Sondaicus?

The above figure was drawn from a stuffed specimen in the British Museum. In colour, shape, and texture of horns, and apparent want of dewlap, it bears some resemblance to the Gaur; but in the skeleton of the Gaur the sacrum consists of *five* vertebræ, and the tail of *nineteen*; while in the skeleton of the Banteng, the sacrum consists of but *four* vertebræ, and the tail of *eighteen*. [Pg 186]

BRITISH DOMESTIC CATTLE.

It does not come within the scope of the present work to give the varieties of Domestic Cattle; for these the reader is referred to the many excellent works already published on the subject. It will be sufficient in this place to notice a few interesting facts—statistical, anecdotal, &c.—in relation to their domestic history.

INFLUENCE OF COLOUR IN BREEDING.

The following remarkable fact, respecting the colour of the offspring being influenced by that of the external objects surrounding the Cow at the time of copulation, is stated by John Boswell, of Balmuto and Kingcaussie, in an essay upon the breeding of Live Stock, communicated to the Highland Society in 1825. He says:—"One of the most intelligent breeders I have ever met with in Scotland, Mr. Mustard, an extensive farmer on Sir James Carnegie's Estate in Angus, told me a singular fact, with regard to what I have now stated.

One of his cows happened to come into season while pasturing on a field which was bounded by that of one of his neighbours, out of which field an Ox jumped, and went with the Cow, until she was brought home to the Bull. The Ox was white with black spots, and horned. Mr. Mustard had not a horned beast in his possession, nor one with any white on it. Nevertheless, the produce of the following spring was a black and white calf, [Pg 187] *with horns.*" Another fact, which shows the great care required in keeping pure this breed — (the Angus doddies) — is related of the Keillor Stock, where, two different seasons, a dairy cow of the Ayrshire breed, red and white, was allowed to pasture with the black doddies. In the first experiment, from pure black Bulls and Cows, there appeared *three* red and white calves; and on the second trial, *two* of the calves were of mixed colours. Since that time care has been taken to have almost every animal on the farm, down to the Pigs and Poultry of a black colour.

INFLUENCE OF THE MALE IN BREEDING.

An ordinary Cow, and a Bull without horns, will produce a calf resembling the male in appearance and character, without horns and without that particular prominence of the transverse apophysis of the frontal bone. The milk of the female from this cross, also, proves the influence of the male: it has the peculiar qualities of the hornless breed — less abundant, containing less whey, but more cream and curd.

GENERATIVE PRECOCITY.

A Mr. Gordon relates the following singular instance of fecundity and early maturity in the Aberdeen Cattle. "On the 25th of Sept., 1805, a calf of five months old, of the small Aberdeenshire breed, happening to be put into an enclosure among other Cattle, admitted a male that was only one year old. In the month of June following, at the age of fourteen months, she brought forth a very fine calf, and in the Summer of 1807, another equally good. The first calf, after working in [Pg 188] the Winter, Spring, and Summer of 1809, was killed in January, 1810, and weighed 6 *cwt.* 3 *qrs.* 16 *lb.* The second was killed December 16, 1810, aged three years six months, and weighed exactly 7 *cwt.*; and on Dec. 30, 1807, the mother, after having brought up these calves, was killed at the age of two years and eight months, and weighed 4 *cwt.* 1 *qr.* the four quarters, sinking the offal."

MILK.

Cows are usually milked three times a day over the greatest part of Scotland, from the time of calving till the milk begins to dry up during the Winter season, when the Cows are for the most part in calf; nor is it found that they suffer by that practice in any degree: and it is the general opinion of all who adopt it, that nearly one third more milk is thus obtained than if they were milked only twice.

A Cow, mentioned by Dr. Anderson in his 'Recreations,' (vol. v, p. 309,) was milked three times a day for ten years running, during the space of nine months, at least, every year; and was never seen, during all that period, but in very excellent order, although she had no other feeding than was given to the rest of the Cows, some of which were very low every winter, when they gave no milk at all.

A farmer of the name of Watkinson had a Cow that, for seventeen years, gave him from ten to twenty quarts of milk every day; was in moderate condition when taken up, six months in fattening, and being then twenty years old, was sold for more than £18. Mr. John Holt, of Walton, in Lancashire, had a healthy Cow-calf presented [Pg 189] to him, whose dam was in her thirty-second year, and could not be said to have been properly out of milk for the preceding fifteen years.

Yorkshire Cows, which are those chiefly used in the London Dairies, give a very great quantity of milk. It is by no means uncommon for them, in the beginning of the Summer, to yield thirty quarts a day; there are rare instances of giving thirty-six quarts; but the average measure may be estimated at twenty-two or twenty-four quarts.

Alderney Cow, after Howitt.

BUTTER.

The Alderney Cow, considering its voracious appetite, yields very little milk; that milk, however, is of an extraordinary excellent quality, and gives more butter than can be obtained from the milk of any other cow. John Lawrence states that an Alderney Cow that had strayed [Pg 190] on the premises of a friend of his, and remained there three weeks, made 19 lbs. of butter each week; and the fact was held so extraordinary, as to be thought worthy of a memorandum in the parish books. The milk of the Alderney Cow fits her for the situation in which she is usually placed, and where the excellence of the article is regarded, and not the expense.

Lord Hampden, of Glynde, had a cow which in the height of the season yielded ten pounds of butter and twelve pounds of cheese every week, and yet her quantity of milk rarely exceeded five gallons per day. The next year the same cow gave nine pounds and a half of butter per week for several weeks, and then for the rest of the summer between eight and nine pounds per week; and until the hard frost set in, seven pounds; and four pounds per week during the frost. Yet as a proof of the quality of the milk, she at no time gave more than five gallons in the day. To this may be added that, "four or five years before, the same person had a fine black Sussex

Cow from Lord Gage, which also gave, in the height of the season, five gallons per day, but no more than five pounds of butter were ever made from it." This is accounted for in a singular way; for there is a common opinion in the east of Sussex, that "the milk of a black cow never gives so much butter as that of a red one."

MR. YOUATT'S PHILOSOPHY OF RABIES, OR MADNESS.

In treating of Rabies, Youatt says:—"When a rabid or mad dog is wandering about, labouring under an irrepressible disposition to bite, he seeks out first of all his own species; but if his road lies by a herd of cattle, he [Pg 191] will attack the nearest to him; and if he meet with much resistance, he will set upon the whole herd, and bite as many as he can.... If the disease is to appear at all, it will be about the expiration of the *fifth week*, although there will be no absolute security in less than the double number of months," After making these remarks, our author reasons himself into the sapient conclusion, that the poison in all rabid animals resides in the saliva, and does not affect any other secretion. "The knowledge that the virus is confined to the saliva," he opines, "will settle a matter that has been the cause of considerable uneasiness. A cow has been observed to be ailing for a day or two, but she has been milked as usual; her milk has been mingled with the rest, and has been used for domestic purposes, as heretofore. She is at length discovered to be rabid. Is the family safe? Can the milk of a rabid cow be drunk with impunity? Yes, perfectly so, for the poison is confined to the saliva. The livers of hundreds of rabid dogs have been eaten in days of ignorance, dressed in all manners of ways, but usually fried as nicely as possible, as a preventive against madness. Some miscreants have sent the flesh of rabid cattle to the market, and *it has been eaten without harm*; and so, although not very pleasant to think about, *the milk of the rabid cow may be drunk without the slightest danger.*"

Is it, indeed, possible for any of the secretions of an animal to be in a healthy state, and fit for human food, after it has had the virus of a rabid dog circulating in its system for at least *five weeks*? Furthermore, is it consistent in Mr. Youatt to call those *miscreants* who send the flesh of rabid cattle to market, when he acknowledges, [Pg 192] in the same breathy that it can be eaten without harm?

According to Mr. Youatt's philosophy, a cow in a rabid state is actually as good as a cow in a healthy state; for its milk may be drunk with impunity—the family is *perfectly safe* who uses it for domestic purposes; and, moreover, *the flesh of rabid cattle may be eaten without harm*. What more can be predicated of cattle in the purest state of health?

STATISTICS.

The number of cattle in Great Britain was estimated by Youatt (1838) at upwards of eight millions. 160,000 head of cattle are annually sold in Smithfield alone, without including calves, or the *dead market*, i.e., the carcases, sent up from various parts of the country. 1,200,000 sheep, 36,000 pigs, and 18,000 calves, are also sent to Smithfield in the course of a year.

A tenth part of the sheep and lambs die annually of disease (more than 4,000,000 perished by the rot alone in the winter of 1829-30), and at least a fifteenth part of the neat cattle are destroyed by inflammatory fever and milk fever, red water, hoose, and diarrhœa.

If a tithe of the sheep and lambs, and a fifteenth of the neat cattle *die of disease*, what proportion are *slaughtered and sent to market in the earlier stages of disease*; and, in fact, in all the stages antecedent to those which are the immediate cause of death?

THE END.

www.ingramcontent.com/pod-product-compliance
Lightning Source LLC
Chambersburg PA
CBHW031627210526
45464CB00004B/1789